Afterglow

Reflecting with the Apollo Astronauts on Their Missions and Lives

Afterglow

Reflecting with the Apollo Astronauts on Their Missions and Lives

DEREK WEBBER

CP
Curtis Press

Derek Webber
Spaceport Associates

ISBN: 978-0-9934002-3-0

© Derek Webber 2017

This trade edition published in 2019 by Curtis Press

Cover design: JudithSDesigns&Creativity, *www.judithsdesign.com*

Distributed in North America by SCB

Visit Curtis Press at *www.curtis-press.com*

Contents

Dedication

To the future space explorers, tourists, and settlers
ensuring the back-up plan for Earth

Foreword

Distance and Destiny: Apollo's Legacy
By Leonard David

It is nearly a half-century ago that the first human cautiously set foot on the time- weathered Moon. The triumphant and transformative achievement was made possible by some 400,000 people that worked together in a unified enterprise to turn a long-held dream into reality. To do so required a blend of government, industry, and academic wherewithal—along with guts, resolve, and risk-taking. As Neil Armstrong and Buzz Aldrin stood there on that magnificent but inhospitable and desolate moonscape, back on Earth an estimated 600 million people witnessed the historic Apollo 11 landing on television.

That first human sojourn to the Moon was propelled by President John F. Kennedy and his seminal "moon speech," given in September 1962 at Rice University in Texas: "But why, some say, the Moon? Why choose this as our goal? And they may well ask why climb the highest mountain? Why, 35 years ago, fly the Atlantic?"

Continuing, Kennedy made clear the why of it all: "We choose to go to the moon in this decade and do the other things, not because they are easy, but because they are hard, because that goal will serve to organize and measure the best of our energies and skills, because that challenge is one that we are willing to accept, one we are unwilling to postpone, and one which we intend to win ..."

That said, my good friend, Buzz Aldrin, reminds me that Kennedy's goal of sending *a* man to the Moon and bringing him back safely to Earth could have played out far differently than what the history books note. Indeed, just one person could have made the trek. Once on the Moon, that individual could have stared out the window, signaled back to Earth *goal achieved*, and then rocket homeward without ever planting a boot print onto the lunar terrain.

There were six Apollo lunar landing missions between 1969 and the close of 1972. Call them the "dusty dozen." They were visitors from Earth that kicked up the lunar dust. Exploration by Apollo moonwalkers was limited: from Apollo 11's humble 2.5 hours to the 22 hours of Apollo 17's campaign of forays, the total time humans spent on the Moon adds up to only 80 hours. That's less than 4 days. But more to the point, and in quite literal terms, the exploration of the Moon—both by robots and humans—is a "barely scratched the surface" endeavor in terms of gathering knowledge about that crater-pocked globe.

Even today, new findings about the Moon are gleaned from investigating the 842 pounds of lunar rocks, core samples, pebbles, sand, and dust brought back by Apollo astronauts from the lunar surface.

Oddly, however, the legacy of Apollo, as momentous as it is, also can be characterized as how *not to do* sustainable human space exploration. There's a growing call for returning to the Moon. But the call this time is to stay. To not only survive there, but thrive there. A cadre of space strategists, engineers, scientists, architects, and others are plotting out a set of themes under the rubric of a Global Exploration Strategy:

- *Human Civilization*: Extend human presence to the Moon to enable eventual settlement.
- *Scientific Knowledge*: Pursue scientific activities that address fundamental questions about the history of Earth, the Solar System, and the Universe—and about our place in them.
- *Exploration Preparation*: Test technologies, systems, flight operations, and exploration techniques to reduce the risks and increase the productivity of future missions to Mars and beyond.
- *Global Partnerships*: Provide a challenging, shared, and peaceful activity that unites nations in pursuit of common objectives.
- *Public Engagement*: A vibrant space exploration program engages the public, motivates students, and helps develop the high-tech workforce that will be required to address the future challenges facing humankind.
- *Economic Expansion*: Expand Earth's economic sphere, and conduct lunar activities with benefits to life on the home planet.

Whatever the future holds regarding the Moon, one thing is for sure. It is far from being what some have labeled as a "been there, done that" world. Furthermore, as this book attests, humankind remains in the afterglow that resulted from Apollo.

The foundation to move forward will be built upon by the legacy of the 24 valiant men described in the following pages. They crossed the vacuum void between the Earth and Moon, carrying with them the spirit of exploring the unknown while setting in motion the destiny that awaits those that follow.

Leonard David
Golden, Colorado, USA

Thanks

I spent nearly 50 years in the UK, before moving to live in the US in 1993. I offer thanks to the late Reg Turnill of the BBC, for keeping me informed through his broadcasts and books in those years during the amazing formative era of the space program. Much later he became a friend, and he even interviewed me about space tourism in Houston in 2002, when he was still carrying out his trade at age 87!

Another great source of knowledge and enthusiasm to me has been the Smithsonian Institution's National Air and Space Museum (NASM) in Washington, DC, to which I offer thanks for its wonderful series of public lectures on astronautics, many of which have informed the text of this book. And to whom I am further indebted for having been chosen, and trained, as one of their volunteer docent guides. Thanks to my fellow docents for accepting me on board and sharing their knowledge so freely. Keep up the good work!

Plaudits to Kim and Sally Poor of Novaspace, creators of the SpaceFest conference series, whose combined professionalism and commitment has resulted in so much of the lunar period's heritage material and space art being saved and shared for future generations to enjoy. And to Robert Pearlman, whose website *Collectspace.com* has been a wonderful forum for continued discussion, and even a source of expertise, for the details of the Moon missions. To the Google Lunar XPRIZE folks, who are working to make possible a new era of lunar exploration, with prizes competed for by teams of "guys in garages" all around the world, thank you for selecting me to be a judge of that competition. It has been a privilege and professional highlight in my life, hopefully in some small way playing a part in getting us back to the Moon. My interest continues to be the future, as well as the past, and so I cherish the support I am given by my team at the Gateway Earth Development Group. We are working to find a much more economical way of exploring the Solar System than the approach described in this book, which was used a half century ago. Keep checking out our website *www.GatewayEarth.space* to follow developments.

With respect to the production of the book itself, I am indebted to Neil Shuttlewood of Curtis Press for his design, attention to detail, and tireless quality control efforts. Arlene Kelly of Vivid Art and Design provided high-quality charts and cover concepts. And Jody Fraser of Transcription Connection performed miracles in transcribing dozens of highly garbled, yet nevertheless precious, microcassette tapes from my collection. I am obliged to Phil Smith for his continued support of my work, and his design expertise.

The veteran space journalist and author Leonard David has kindly provided the Foreword, and I am immensely grateful for this gesture of support. I thank my wife Sarah Fisher, who has again provided me with the absolutely necessary positive vibes while I have been preparing the book, and also for proof-reading my manuscript draft. Any errors in text or attributions remaining are my own, and I would appreciate reader feedback so that I can make the necessary corrections.

Finally, I must thank the Moon travelers themselves, for risking their lives in the first place, for thereafter never tiring of telling the tale, and finally for encouraging me, when the occasion presented itself, to keep on trying to build a future where space tourism enables a new era of space exploration.

Derek Webber
Damariscotta, Maine, USA
July 2017
DWspace@aol.com
www.SpaceportAssociates.com

Abbreviations

ACA	Attitude Control Assembly
ALSEP	Apollo Lunar Surface Experimental Package
ALT	Approach and Landing tests (of Space Shuttle)
AMU	Astronaut Maneuvering Unit
ASTP	Apollo–Soyuz Test Project
ATC	Air Traffic Control
ATC	American Television and Communications, Corp
BIG	Biological Isolation Garment
BIS	British Interplanetary Society
B.Sc.	Bachelor of Science
CAPCOM	Capsule Communicator
CDR	Commander
CIA	Central Intelligence Agency
CEO	Chief Executive Officer
CM	Command Module
CMP	Command Module Pilot
COMSTAC	Commercial Space Transportation Advisory Committee (of the FAA)
COO	Chief Operating Officer
CSM	Command and Service Module (of Apollo spacecraft)
DC-X	Delta Clipper Experimental
D.Phil.	Doctor of Philosophy
DVD	Digital Video (or Versatile) Disk
ESA	European Space Agency
ESP	Extra Sensory Perception
EVA	Extra Vehicular Activity (spacewalk)
FAA	Federal Aviation Authority
G	The force of gravity
GDP	Gross Domestic Product
GEDG	Gateway Earth Development Group
GEO	Geostationary Orbit
GLXP	Google Lunar XPRIZE
GT	Gemini/Titan spacecraft and launch vehicle combination
HDTV	High-Definition Television
HiRISE	High-Resolution Imaging Science Experiment (of Mars Global Surveyor)
IMAX	Image Maximum (cinematic high definition movie system)
ISDC	International Space Development Conference (of the NSS)
ISRU	In-situ Resource Utilization (robotic device to process extra-terrestrial materials)

ISS	International Space Station
ISU	International Space University
JFK	John F. Kennedy
JSC	Johnson Manned Spaceflight Center (NASA-Texas)
KSC	Kennedy Space Center (NASA-Florida)
LEM	Lunar Excursion Module (same as LM)
LEO	Low Earth Orbit
LLTV	Lunar Landing Training, or Test Vehicle
LM	Lunar Module (same as LEM)
LMP	Lunar Module Pilot
LP	Long Playing (vinyl) record
LRL	Lunar Receiving Laboratory
LRO	Lunar Reconnaissance Orbiter
MA	Mercury/Atlas spacecraft and launch vehicle combination
MECO	Main Engine Cut-Off
MET	Modular Equipment Transporter
MIT	Massachusetts Institute of Technology
MQF	Mobile Quarantine Facility
MR	Mercury/Redstone spacecraft and launch vehicle combination
MS	Master of Science
MSFC	Marshall Space Flight Center (NASA-Alabama)
NAC	NASA Advisory Council
NASA	National Aeronautics and Space Administration
NASM	National Air and Space Museum of the Smithsonian Institution
NSS	National Space Society
PAX River	USN Test Pilot School, Patuxent River, Maryland
PD	Powered Descent
Ph.D.	Doctor of Philosophy
POW	Prisoner of War
PR	Public Relations
SIM-Bay	Scientific Instrument Module—location on Apollo CSM
SpaceX	Space Exploration Technologies corporation
STS	Space Transportation System (Space Shuttle)
TEI	Trans-Earth Injection
TLI	Trans-Lunar Injection
UN	United Nations
USSR	Union of Soviet Socialist Republics (the Soviet Union)
VAB	Vehicle Assembly Building
VIP	Very Important Person
VP	Vice President
WW	World War 1 or 2
WWW	World Wide Web

CHAPTER 1

Introduction

It was the Golden Age. It started in 1961, lasted just over a decade, and by the time it ended 24 humans had journeyed the vast distance to the Moon and back (Figure 1.1). Before it started, no humans had been in space at all. It was an amazing period for humanity to witness. These 24 humans have still, half a century later, been the only ones to see at first-hand what our entire planet looks like, slowly spinning without any visible means of suspension in the vastness of space. That experience must change a person, surely?

It has been said, quite correctly, that the astronauts who crewed the missions were merely "the point of the arrow," metaphorically as well as physically as they were strapped in high up on their rockets, so why concentrate on them? There were nearly a half million people, all over the United States, who provided these travelers with means of transport and life support, and most of them have never recounted *their* individual experience in book form. We may be sure, however, that they certainly talked about it with friends and family, and we must therefore rely on those accounts in the oral tradition. This was such a significant and rapid series of events, and it was given so much priority by the White House and Congress and a supporting press, that no matter what was the space-related job of each of those 400,000, the progress of the space program was never very far from being the topic of daily conversations (and that was true for the rest of the population also). Ask any of them who are still around even today, and they will tell you their tales—they really enjoy talking about it. After all, it was considered an almost existential crisis at the time. It was a war without having to kill an enemy, and this particular aspect of the Cold War, namely the space race between the Soviet Union and the USA to reach the Moon, was won decisively by the US when the crew of Apollo 11 returned safely to Earth on July 24, 1969. This was an American success story. In many ways it defined a generation.

We focus, then, on the Moon voyagers themselves, and this account is only about US space explorers. The Soviets had their own heroes, and in the early years of the space race, they were indeed in the lead. But in this book, we are looking for a certain kind of completeness to the US experience, and are therefore focusing on *all 24 of the American astronauts who made the journey to the Moon*. We shall find that, although they had a lot in common, they displayed very different personalities—some being amazingly different—which colored the

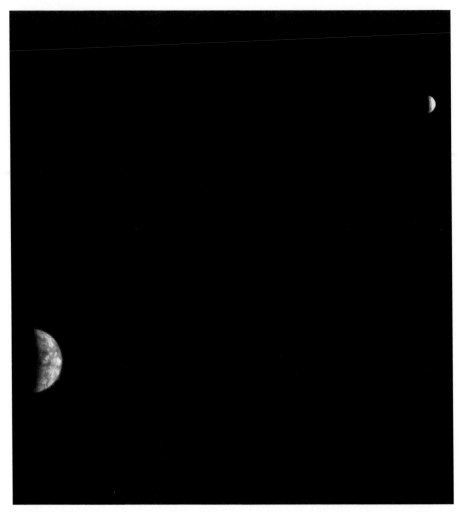

Figure 1.1. The vast distance between Earth and the Moon that was traveled by 24 humans between 1968 and 1972. This image was produced on October 3, 2007, by a telescopic camera called HiRISE looking back home from the Mars Reconnaissance Orbiter spacecraft circling Mars. The full distance between the two celestial bodies is thirty times the Earth's diameter.
Credit: NASA

lives they subsequently led upon returning from the Moon. Furthermore, even amongst the 24 Moon travelers, they each had distinctly different experiences during their missions. For example, 12 of them walked on the Moon, 6 of them experienced flying solo around the Moon, 3 floated outside their capsule while in deep space near the Moon, and 3 of them even made the perilous lunar journey twice. Who were these first emissaries of the Earth, who ventured out to the Moon, and how did the journey affect each one of them? We shall provide answers based upon their own words, both spoken at the time, and over the 50 years since their missions, and on interactions the author has had with the majority of them, discussing such things as their perspective on the future of space exploration. In the summer of 2019 it will be the Golden Anniversary of the successful outcome of that Golden Age of spaceflight, an appropriate time to reflect on this band of pioneering explorers. In some respects, there are parallels with looking for the records of the crews of Columbus's ships the *Niña*, *Pinta*, and *Santa María*, or of the reflections of the 29 adult passengers on the *Mayflower*. These crews who went to the Moon also reflect a new beginning. In the case of the *Mayflower*'s pilgrims, it is possible today to track their progeny all the way from 1620, when by now there are about 35 million descendants of that defining voyage. Most of the 24 Moon travelers also had children and grandchildren whose own descendants will provide an ongoing connection with these first lunar explorers, whose pioneering stories are collected herein. We may even suppose that some of these descendants will continue the tradition by being space travelers themselves, and possibly ultimately becoming settlers elsewhere in the solar system—so it is well worth documenting their origins, as an atavistic touchstone for future generations. Many of these Moon travelers have written complete books focused only on their own missions (all of which are included for you in the reference section of this book). So, clearly, we must in this one, which covers all 24 of the lunar astronauts, provide a distilled essence for each of them, which reflects their thoughts and their significant contributions both at the time of their missions, and later. I write as someone professionally involved in the space program initially in the UK, and subsequently in the US, and as a former volunteer docent guide for the Smithsonian Institution's National Air and Space Museum (NASM). I am especially interested in the future of space travel and exploration, and what we might learn from their unique perspective.

We have pointed out that we are talking about a mere decade—the 1960s. Even in a professional life of say 50 years, a decade represents a small part. Of course in a full lifetime it represents not much more that 10% of one's span. So, it is not surprising that for many of the guys—they were all male back then—there is perhaps some resentment that those 10 years have come to define them as people. They may even have fought hard to prevent that from happening. But none of them have succeeded. It will always be the first thing in their obituary notice: "Traveled to the Moon on Apollo." But maybe in this book, we can round out the picture somewhat, and provide a fuller perspective on their lives.

We can also attempt to draw some conclusions at this remove about the way the missions themselves were conducted. How successful were they? What do we learn about the recruitment processes? What about the training that was used? In some respects, this is all part of the legacy of Apollo—passing on what we learned, and even felt, during that "Golden Age" when we first went to the Moon. Things were done in such a hurry, in order to fulfill the President's challenge to complete the Moon landing "before this decade is out," that not everything was neatly wrapped up after the program ended. So we can do some of that completion work now. There is no better way to demonstrate the mindset when the project ended, with a new focus on subsequent missions including the development of the Space Shuttle, than to note the surprising fact that *there was never a single photograph taken which showed all 24 men who went to the Moon.* No one even thought to take a group photo of the 12 men who walked on its surface. In an age of digital phones, such as we have now, it is hard to understand such an omission. Part of the explanation is cultural—we need to appreciate the period scene.

We have come so far so quickly in the years following the lunar landings, that it is sometimes hard to remember just how things were. Take TV, for instance. Why were the first pictures we received on the Moon so fuzzy? Back at the time, they were not perceived in that way—in fact all the world was amazed that we were able to get TV from the Moon at all! TV camera equipment was large and clunky and used up lots of power, so special equipment had to be designed for use on the Apollo missions. In the 1960s, viewers watched *Perry Mason* in black and white. Network TV broadcasts switched to color in 1972, which was the year when sales of color TV sets first surpassed those of black and white sets, after the Moon missions ended. What about cameras? Cameras all used film in the 1960s, whether for still or moving images, and therefore there was a limitation on the number of shots that could be taken. Digital cameras did not become common among consumers until the late 1990s. There were no mobile phones. The first Motorola mockup was demonstrated in 1973. It weighed 1 kg, had 30 minutes of talk time, and took 10 hours to recharge. It was not until the 1970s that the first battery-operated hand-held electronic calculators were sold. Through much of the 1960s the most common way for engineers to do their work was by using slide rules and paper and pencil, augmented by mechanical calculators when more precision was needed. The sound of these large clunky machines, added to that of typewriters, provided a constant background noise-scape in engineering offices of the period. Personal computers did not come on the scene until the late 1970s. Of course the Internet and World Wide Web came much later (in the 1990s). There were no gaming consoles; no Google! Music was available only on LP vinyl records, or tape cassettes. This was a world before credit cards; before ATM cash machines. Imagine that! You had to go into a bank, in those hours when they were open, and fill out a withdrawal form to get out your own money! It was also the case that the role of women in society was still very restricted. None were able to fly in military combat jets, for example, which automatically ruled them out from meeting the early astronaut entry requirements at the time. So, perhaps this gives you a better appreciation of the challenge that was posed by President John F. Kennedy, when, on May 25th, 1961, after only 15 minutes of accumulated US human spaceflight experience, he announced to Congress and the world that: "I believe that this nation should commit itself to achieving the goal, before this decade is out, of landing a man on the Moon and returning him safely to Earth."

We are going to tell the story of the 24 men who ventured to the Moon in that era which today seems so primitive technologically. We shall cover all of their spaceflights, and have decided on what's probably the best, and most interesting, way to provide the information. In this book the 24 lunar travelers are introduced *in the order in which they took their first spaceflights*, even when that meant on Mercury or Gemini flights, and not therefore in the simple chronological order of the nine Apollo Moon missions. Thus, the first one we shall encounter will be **Al Shepard**, even though he did not fly to the Moon until five earlier lunar flights had taken place (he flew on Apollo 14). Also, to avoid making the text too repetitive or lacking in vibrancy, most of the common data features for the two dozen Moon travelers is contained in standard format, mini-bio appendices. In this way, we can use the main text to convey more about their personalities and the accompanying excitement of the Moon missions. Section I of this book will describe the Golden Age—covering selection and training (through Mercury and Gemini), followed by the Apollo Moon missions themselves in Chapter 4, while Section II considers what has happened since that time, including their ideas about the future of space exploration, up until June 2017. We include a comprehensive reference source section, so that we can link the various events and quotations to their appropriate context—the references include some wonderful DVDs,[1.19] which provide great period footage of the missions. Note that we have employed throughout the convention of capitalizing the Moon, since in this book there is no doubt that we are talking about the Earth's celestial neighbor as a named astronomical object. And, furthermore, to make it easier to distinguish those who made the journey from all the other astronauts named in the narrative, we have adopted the convention of **emboldening** their names, and their *italicized* quotations, whenever they occur in the main text.

I said that there was no single group photo of the entire cadre of the 24 Moon travelers, but we close this introduction with what is generally regarded as the best, most complete, grouping of Apollo astronauts in a photo (Figure 1.2). In it we do see 17 of the 24 Moon travelers, including 9 of the 12 men who walked on the Moon. The following lunar astronauts, however, were not present for various reasons when the photograph was taken, on August 21, 1976, at Johnson Space Center (JSC), Houston Texas: **Bean**, **Borman**, **Haise**, **Mattingly**, **Swigert**, **Young**. It's too late now, for sure, to fix that omission.

Figure 1.2. A group of astronauts reunite on August 21, 1976, at JSC, Houston, Texas, four years after the final Moon mission. Included in the image are the following 17 Moon travelers: **Shepard**, **Cernan**, **Anders**, **Duke**, **Lovell**, **Evans**, **Scott**, **Armstrong**, **Worden**, **Roosa**, **Stafford**, **Mitchell**, **Gordon**, **Conrad**, **Aldrin**, **Irwin**, and **Collins**. By the time you have read this book, I guarantee you will be able to tell which is which! Also included in the gathering were five astronauts who did *not* make the Moon journey (Schweickart, Schirra, Cooper, Pogue, and Cunningham).
Credit: NASA

SECTION I

The Way It Was

"Tis likely enough that there may be means invented of journeying to the Moon; and how happy they shall be that are first successful in this attempt."

Dr. John Wilkins

A Discourse concerning a New World and Another Planet, 1640

CHAPTER 2

Getting Started

THE FIRST TWO OF THE EVENTUAL MOON TRAVELERS GET INTO SPACE

In this chapter, we shall be reviewing a 6-year period at the very beginning of the US space program, during which 2 of the eventual 24 lunar explorers had their first spaceflights. The period therefore starts in April 1959 with the recruitment of the Mercury astronauts, and includes the time between **Alan Shepard**'s first flight (in the first Mercury capsule) and **John Young**'s first flight (on the first Gemini spacecraft), in 1965. During this period, the first four groups of astronauts, from which 15 of the Moon travelers would be chosen, were all selected. The remaining 9 would not enter the program until Group 5 was recruited, in 1966, and we shall give them due attention in Chapter 3. The succession of astronaut entry groups and their respective entry requirements, as they changed during the "Golden Age," are provided in Appendix B. Even as this book is being written, NASA continues to recruit astronauts. More than 18,300 people applied for the 2017 astronaut class. The previous class to be recruited—in June 2013, consisting of eight newcomers calling themselves "The Eight-balls"—was the 21st intake of NASA astronauts, and consisted of four men and four women, selected from 6,300 applicants. Clearly, being an astronaut is considered a very special job, and arguably this is due to the way in which the public were introduced to the first group to be recruited. Nobody was more surprised than the astronauts themselves at the public response when it was duly announced, on April 9, 1959, that they had become the first Americans to attempt to go into space.

Group 1 were known as "The Mercury Seven." We can see them in Figure 2.1 with Dr. Wernher von Braun, the head of the team of German rocket engineers that came to the US at the end of WW II, whose experience and drive fueled much of the US space program during the "Golden Age." They provided their own combined account of that early period,[21.6] and Tom Wolfe's treatment[1.20] became a great book and movie.

The Mercury Seven had been selected following a tough recruitment process which began on January 5th, 1959, when NASA finalized its selection criteria. President Eisenhower had, in December 1958, decided that the potential astronauts would be selected from the ranks of military test pilots. The full selection

Figure 2.1. The Mercury Seven (Grissom, Schirra, **Shepard**, Glenn, Carpenter, Cooper, Slayton) with von Braun in 1959 alongside his Redstone rocket. **Shepard** would be the first to fly the booster, and as it would turn out, the only one of them to eventually reach the Moon.
Credit: NASA

criteria for each of the astronaut groups are listed in Appendix B for easy reference, but the key points were that they all, in addition to being military test pilots, needed to be under 40, less than 5 feet 11 inches (due to the small size of the spacecraft that was being designed), be in excellent physical shape, and have earned a bachelor's degree. Of 508 test pilots at the time, 110 men met the basic criteria (no women at that time had been given the opportunity to be military test pilots), and were invited to Washington in February 1959 to be given a briefing on the "Man in Space" project. Subsequently, those who had volunteered were given a series of written tests and psychiatric examinations, and 32 remained to undergo the physical tests at the Lovelace Clinic in Albuquerque, and psychological tests at Wright-Patterson Air Force Base, during March 1959. On April 2, seven men were finally chosen to be the nation's "Mercury Astronauts."[1.1]

This method of selection almost guaranteed that those chosen would be fiercely competitive. They had so much in common, including a need to see the

US emerge victorious from the Cold War, which was an era of mutual distrust and fear of nuclear annihilation which operated between 1945 and 1991, only ending when the Soviet Union itself disintegrated. They were all born around 1930, which was just 3 years after Lindbergh flew the Atlantic, and therefore they were fully aware of World War II (in which some of their parents had participated), and as military jet pilots they had flown in the Korean War, and must have been contemplating the possibilities of a Vietnam air war at the time of their recruitment as astronauts, because the US already had advisors training the South Vietnamese, with the first US casualties having emerged in 1959.

The "jet age" had begun at the end of the Second World War in 1945, and adventurous young folk grew up wanting to fly ever higher and faster. In 1947, a 24-year-old Chuck Yeager had taken his Bell X-1 through the "sound barrier" as the USAF developed rocket-powered X-Planes and tested them over the Mojave Desert from Edwards Air Force Base in California. In fact, there was every expectation that this approach to flying faster and higher in aircraft such as the X-1 would ultimately lead to reaching space. So, the Mercury Program, when it was announced, came from outside of the mainstream for military officers. These young candidates had to weigh up which was going to be the successful route to space—the X-plane route or the "capsule on a rocket" approach. On October 4, 1957, the Soviet Union launched the world's first artificial satellite, Sputnik 1, followed by a succession of ever more capable craft, including one carrying the dog *Laika* on November 3, 1957, and had conducted early Moon probe missions. This gave young US aviators concerns. It was January 31, 1958 before the US was able to launch its own satellite, Explorer 1. If dogs could be flown, surely men would come next. In that Cold War era, it was thought essential that any capabilities of the Soviet Union could be matched by the US, and ideally bettered. The US was clearly behind in space capabilities, and the competition which became known as the Space Race would be played out using rockets from launch pads, not using aircraft from runways, such as those being tested at Mojave.

Soon after the Mercury Seven were selected, a new young president, John F. Kennedy, was elected in November 1960, taking office in January 1961. It would be JFK who would signal the importance of winning this space race in an unambiguous way, leaving none of the astronauts with any residual doubt about the importance of the tasks they were taking on, or indeed of the way they had to do it. Kennedy started it, supported by his Vice President Lyndon Johnson, but he was never able to witness the ultimate success, because he was assassinated in Dallas in November 1963. He did, however, enjoy the early achievements of the US space program, and, as a former WWII Naval Lieutenant, spending time talking with his fellow military officers, the Mercury astronauts, whenever he could create the opportunity.

President Eisenhower had set up NASA in July 1958 as a response to the launch into orbit of Sputnik 1, and as we have mentioned, planners were clearly expecting that, following those early animal flights, there would be Soviet men in space. The recruitment process for American astronauts was already set in place by January 1959. Kennedy had seen in 1957 the shock effect of Sputnik on the American public, and the highly positive way in which the Mercury astronauts were received in April 1959, so he very quickly embraced the space program as a way of demonstrating superior technologies to the world, and the Soviets in particular. The first US manned spaceflight would be by **Alan Shepard** on May 5, 1961, just 4 months after Kennedy entered the White House. Only three weeks later, he would announce the Moon objective. Kennedy made two important speeches about the space program during his brief three years as President, and together they laid out the plan, and ensured its funding. Since these words were so important to the future development of US, and indeed world, space technology and operations, let's review exactly what he said in the shaded box below, to understand his motivations.

THE MOON DIRECTIVE

The first speech was on May 25, 1961 before a joint session of Congress, and in it he set the United States on a course to the Moon. The speech itself was about a number of issues, which he described as "Urgent National Needs," including such non-space things as a manpower training program, increased radio and TV broadcasting to South-East Asia, more funds and resources for the military, for nuclear fallout shelters, and support for disarmament talks in Geneva, requesting Congressional support and funding. It lasted a full 46 minutes, of which the Moon part, which was at the end, lasted only 8 minutes. The language of this Moon part is, however, well-known, and is often quoted in management courses as an example of the perfect way to provide a directive: "I believe that this nation should commit itself to achieving the goal, before this decade is out, of landing a man on the Moon and returning him safely to Earth." Thus addressing these questions: Where are we going? The Moon. With what? A man. When are we going? Before this decade is out. But it is also worth noting some other points JFK made during these 8 minutes, because they give us context. "It is a most important decision we make as a nation. But all of you have lived through the last four years and have seen the significance of space and the adventures in space, and no one can predict with certainty what the ultimate meaning will be of the mastery of space." He is here referring to the period since Sputnik 1 was launched. "If we are to win the battle that is now going on around the world between freedom and tyranny, the dramatic achievements in space which occurred in recent weeks should have made clear to us all, as did the Sputnik in 1957, the impact of this adventure on the minds of men everywhere, who are attempting to make a determination of which road they should take." Here he is referring to the spaceflights of Yuri Gagarin (April 12, 1961) and **Alan Shepard** (May 5, 1961), and the choice for nations which were in a tug-of-war between a capitalist or a communist doctrine. Kennedy of course realized the enormous political risk he was taking. He makes this clear: "Recognizing the head start obtained by the Soviets with their large rocket engines, which gives them many months of lead-time, and recognizing the likelihood that they will exploit this lead for some time to come in still more impressive successes, we nevertheless are required to make new efforts on our own. For while we cannot guarantee that we shall one day be first, we can guarantee that any failure to make this effort will make us last. We take an additional risk by making it in full view of the world, but as shown by the feat of astronaut **Shepard**, this very risk enhances our stature when we are successful. We go into space because whatever mankind must undertake, free men must fully share." And then, of course, he asked for the money: "Let it be clear that I am asking the Congress and the country to accept a firm commitment to a new course of action—an estimated seven to nine billion dollars additional over the next five years. If we are to go only half way, or reduce our sights in the face of difficulty, in my judgment it would be better not to go at all." This was the amazing clarion call which the astronauts were called upon to help fulfill.

Kennedy made his second Moon speech on September 12, 1962 at Rice University Stadium in Texas, after having visited the new Manned Space Center in Houston. By this time, **Shepard** had been followed into space by Gus Grissom, John Glenn, and Scott Carpenter. The speech was an inspiring underlining of the commitment to the Moon mission, containing wonderful phrases: "This generation does not intend to founder in the backwash of the coming age of space," he stated, "We set sail on this new sea because there is new knowledge to be gained, and new rights to be won, and they must be won and used for the progress of all people." "We choose to go to the Moon in this decade, and do the other things, not because they are easy, but because they are hard, because that goal will serve to organize and measure the best of our energies and skills." He returned to the subject of budget: "To be sure, all this costs us a good deal of money. The budget now stands at $5,400 million a year—a staggering sum, though somewhat less than we pay for cigarettes and cigars every year. Space expenditures will soon rise some more, from 40 cents per person per week to more than 50 cents a week." Only a year later, on another visit to Texas, Kennedy was shot dead. He had, however, marshaled the resolve of the US people to embark on a journey quite literally into the unknown. The target was the Moon and the timescale to establish the feat was just $8\frac{1}{2}$ years from start to finish. Also, we should remember for future reference, the task as defined would be deemed completed after only one successful Moon landing and return to Earth.

So, Kennedy was the spur. But work was already underway before he took office. Let's look now at those beginnings—the early program leading up to **Shepard**'s flight in the Mercury capsule *Freedom Seven*—which however inauspicious were nevertheless convincing enough to the new President to allow him to take on the tremendous political risk of the Moon mission. When the Mercury Seven came on board, they had a great deal to learn. As did everybody else on the program, including the ground control support team. A curriculum was developed in real time to train the astronauts about all aspects of their mission, including spacesuit design, medical factors, pre-launch, launch, orbital theory, navigation, zero-g, re-entry, survival training, landing, and recovery. Later on, during the Gemini and Apollo missions, other aspects were added—rendezvous, docking, extra-vehicular activities (EVAs), and geology and quarantine procedures. The astronauts were subjected to testing in centrifuges—sometimes up to 16g[3.1]—flying zero-g profiles in specialized aircraft, multi-axis disorientation devices, and an array of different simulators. Later on, they also used a large water tank—the buoyancy chamber—to provide zero-g experience and training. In the course of the exhaustive testing of failure modes during simulator training, there was much writing, and re-writing, of check lists. They were all test pilots so this type of work was familiar to them. They were also able to maintain their flight readiness by using the astronauts' own fleet of T-38 Talon jet trainers. Meanwhile, of course, they had to move their families to the new homes they had built in close proximity to the new Johnson Space Flight Center in the Clearlake area of Houston, Texas, and their children had to be introduced into local schools. Fortunately, being in close mutual proximity provided advantages for the wives, and since increasing numbers of astronauts were moving into the district, as far as the children were concerned, it was No Big Deal to have an astronaut as a father, and life was therefore not unduly disrupted for them. Being the focus of the nation's attention had brought the Mercury Seven some benefits following their introduction to the public. In particular, they were encouraged by NASA to sign an exclusive deal with *Life* magazine, so that the stories of the astronauts and their families could be covered in a controllable way. Eventually, over 30 cover stories about the space program and the astronauts would appear in *Life* magazine in the decade between 1959 and 1969. The astronauts began to experience the kind of life that celebrities led, with at least their financial concerns being removed from the list of matters to be worried about. With all the 24 astronauts who eventually made the Moon trip, including **Shepard**, summary biographical data can be found, presented in alphabetical order, in Appendix A.

Alan Bartlett Shepard, although successfully competing with his fellow Mercury astronauts in order to be selected for the coveted first US man in space assignment, was beaten into space on April 12, 1961, by the Soviet Union's Yuri Gagarin. This was not only a month earlier, but technically it had been a much more ambitious feat, since Gagarin went into orbit, whilst the first two Mercury flights were only sub-orbital lobs. President Kennedy was well aware of this, as is clear from his comments after **Shepard** returned safely to Earth. The US was starting from behind in the Space Race. **Shepard** himself refers to this:[21.2]

Shepard:	*"We flew another unmanned mission before Gagarin flew, then his flight, then mine, so it was really touch-and-go there. If they had put me [instead] in that unmanned mission, we would actually have flown first."*

Questioner: "In retrospect it doesn't seem that important, but at the time I guess it was?"

Shepard: *"Oh, very important; absolutely, absolutely."*

Shepard was a Navy pilot who had served on aircraft carriers in the Mediterranean before becoming an instructor at the Navy's test pilot school at Naval Air Station Patuxent River ("Pax River") in Maryland, leading to his being recruited onto the space program. His father was a career Army officer, and it seems that for much of his professional life, **Shepard** was in effect trying to live up to the need to satisfy his wishes. Indeed, we learned later that, even after becoming the first American in space, his father's reaction was still not favorable. He regarded his son's entry into the space program as a diversion in his erstwhile career as a Navy officer. It was only after his eventual return from walking on the Moon that his father seemed able to accept the amazing achievements of his son. **Shepard** was to eventually retire from the Navy as a Rear Admiral. But certainly at the start, neither **Shepard** nor any of his Mercury colleagues had any expectation of the Moon as a destination. His first flight into space was in hindsight quite a modest affair. He flew in his Mercury capsule, which he had named *"Freedom Seven,"* on top of a Redstone booster for a 15-minute flight down the Atlantic Missile Range from Cape Canaveral, before landing by parachute in the Atlantic Ocean and being recovered by the US Navy. Since the start of the program, there had been around 800 changes incorporated to "man-rate" the former von Braun Redstone missile, (usually related to improved redundancy of systems), to make it ready for manned flight. Although the flight itself was brief, no modern reader should underestimate the courage that was required to undertake **Shepard**'s task that day. Everything was unknown. Would the rocket blow up, as many of them did in that era? Could an astronaut survive the launch loads? Would the spacecraft hold up in the vacuum of space? Would the spacesuit work? How about communications? And then of course, there were worries about re-entry, the heatshield, parachute deployment, and recovery.

Back in the UK, the author, as a 16-year-old boy, was intent on following the radio news to track the mission. The best source of information was the BBC with Reginald Turnill as special space correspondent giving reports direct from the space center at Cape Canaveral. You could also follow the development of the space race in publications such as the French magazine *Paris Match*, which was particularly good at printing good color photographs. Later there were View-Master 3-D reels which documented the US space missions for eager young space buffs. There was also news of the Soviet missions, and we followed the flights of the Vostok cosmonauts from Gagarin onwards. But it was the US program which provided the best imagery and information for us in the UK, and indeed for most of the world. In a very real sense, the whole world was being taken along for the ride as these first space explorers went off into the unknown. In Figure 2.2 we observe **Shepard** getting ready for his Mercury Redstone mission.

Figure 2.2. Shepard is suited up before his history making flight in the Mercury spacecraft *Freedom Seven*.
Credit: NASA

Shepard was frustrated by the various launch holds, leading to him calling out:

"Let's light this candle!"

over the intercom.[21.7] There were thousands of folk watching from Cocoa Beach, willing him up into space, as the vehicle finally took off (Figure 2.3). At

Figure 2.3. *Freedom Seven*, carrying **Alan Shepard**, heads into space, at 9:34 AM EST on May 5, 1961.
Credit: NASA

the time, every word spoken by an astronaut in space, however mundane, was recorded for posterity,[21.5,21.6] and much of the communication link was broadcast in real time over the PA systems installed for the purpose:

"Roger, liftoff and the clock is started .... This is Freedom Seven, the fuel is go. 1.2 g, cabin at 14psi, oxygen is go .... On periscope. What a beautiful view."

Then 15 minutes later:

"3, 6, 9, OK, OK, OK"

(calling off g-stresses, which would eventually reach 11g). Then, as the parachutes were deployed:

"Roger. The drogue green at 21,000. Good drogue. The drogue is good. Seventy percent auto, nine zero percent manual. Oxygen is still OK."

Figure 2.4. Naval aviator turned hero. A lapel pin recording the event.
Credit: Author's collection

Then he landed and was picked up by the carrier *USS Lake Champlain*, which, together with eight destroyers and a radar-tracking vessel, provided the recovery fleet for the mission. That's quite a reception party for a 15-minute lob. Imagine how many naval crewmen were individually involved, and had trained for that day, and remember their feelings still. Shortly after his arrival on the carrier, he received a telephone call direct from President Kennedy, offering his congratulations. The medics checked him out and found him to be in good shape following his flight.

After his successful recovery downrange, his transformation into national hero was cemented through such things as lapel pins (Figure 2.4), visits to Congress and the White House (Figure 2.5), being a celebrity guest on TV chat shows, and having his photo on the cover of *Life* magazine and even postage stamps (not, at this stage on US stamps, but on stamps from such places as Togoland—it was only after his death that he made it onto a US stamp). He even made it to London to pick up a Silver Award for the Mercury program from the British Interplanetary Society, and there were long-playing vinyl records made to record the 15-minute feat. What was the title of the LP? Check out Figure 2.6 for a 1961 best-selling hit record!

Each succeeding Mercury mission to some degree pushed the envelope of possibilities. Gus Grissom flew next, on another sub-orbital trajectory, then began to focus his attention on cockpit design for the new two-man program Gemini, eventually flying on its first flight with **John Young** as his co-pilot. Then, John Glenn was the first American to fly into orbit (3 orbits), he subsequently became a Senator, and many years later flew on a Space Shuttle mission. Scott Carpenter followed and also flew for three orbits, then retired from NASA and became an aquanaut, exploring the depths of the ocean instead of space. Wally Schirra subsequently flew 6 orbits on Mercury, then flew on Gemini with **Tom Stafford**, and eventually flew on Apollo with Donn

Figure 2.5. The **Shepards** meet the Kennedys and Lyndon Johnson at the White House.
Credit: NASA

Eisele and Walt Cunningham, though only in Earth orbit. Gordo Cooper was last, and flew for 22 orbits on Mercury, wrapping up that program, then flew with **Pete Conrad** on Gemini. Deke Slayton was never able to make a Mercury flight, being grounded for a heart murmur, but finally flew after the Moon-landing era, on the joint US/Soviet mission ASTP (Apollo–Soyuz Test Project), with **Tom Stafford** and Vance Brand. These one-man Mercury flights had provided a foot-hold for the US in the realm of manned spaceflight.

Slayton became Chief of the Astronaut Office when he was grounded for heart-related health reasons around the time of Glenn's Mercury flight, and was therefore placed in charge of crew selection, establishing the crew-rotation process for new astronauts.[1.15] The process generally required that a new "rookie" astronaut would first be assigned as a member of a support crew for a mission, then as a member of a back-up crew for another flight, before finally

Figure 2.6. It is interesting to note that **Shepard**'s name did not even appear on this celebratory LP record's cover. Everyone at the time knew exactly who "America's First Astronaut" was.
Credit: Author's collection

joining a prime crew. **Shepard** was himself also later grounded for medical reasons a rare condition known as Menière's disease which affected his balance—and subsequently worked with Slayton to manage the astronaut program leading to the Moon landings. At that time, Slayton became Director, and **Shepard** became Chief, of the Astronaut Office. They were a formidable team. As reported by Wolfe,[1.20] **Shepard** was known amongst the astronaut office as a person with two distinct personality traits. On the one hand, and certainly while establishing himself among the Mercury Seven, he could be fun-loving; however, and especially after he became grounded and was responsible for managing astronaut affairs and training, he could be very firm and remote. His secretary at the time used to have two different photos of her boss, and would stick the appropriate one on the door depending on the mood he was in, as a warning to those requesting to meet with him. One showed "Smilin' Al," and the other one showed "The Icy Commander." Astronauts learned not to come with problems on the wrong day.[1.20] We shall encounter **Al Shepard** later in this book after he regained flight status in time for a Moon trip, and when the author had a chance to meet him,[21.4] but now we move on. Look to Appendix C for a summary chart of all the key missions—we have noted that the last Mercury flight was carried out by Gordo Cooper on May 15, 1963, and at that time it was already recognized that more astronauts were going to be needed to fulfil Kennedy's goal. Group 2 was recruited in June 1962, and Group 3 in October 1963. The first two-man Gemini mission would be on

Figure 2.7. Group 2, June 1962, "The New Nine." *Top row*: **Armstrong**, **Stafford**, and White. *Middle row*: **Borman**, **Lovell**, and McDivitt. *Front row*: See, **Conrad**, and **Young**.
Credit: NASA

March 23, 1965, and so the new recruits were quickly inducted into the training programs that Slayton and **Shepard** were managing.

Group 2 (Figure 2.7), were known collectively as "The New Nine," and six of the nine (**Lovell**, **Stafford**, **Conrad**, **Borman**, **Armstrong**, and **Young**) would eventually become travelers to the Moon. Of the remaining three, See would lose his life in a training plane crash, White would become the first American to "space-walk," then would die in a fire during an Apollo test, and McDivitt would command two important missions but never leave Earth orbit.

John Young would be the first of this group to get into space, so we shall learn about him and his first flight later in this chapter, after noting the arrival of the next group of astronauts.

Group 3 (Figure 2.8) were known as "The Fourteen," and seven of them (**Collins**, **Gordon**, **Scott**, **Aldrin**, **Anders**, **Bean**, and **Cernan**) would eventually become Moon travelers. Of the remainder, Freeman, Williams, Bassett, and Chaffee would die in training before ever making a spaceflight. Cunningham, Eisele, and Schweickart did each make one spaceflight, but only in Earth orbit.

Figure 2.8. Group 3 astronauts, October 1963, "The Fourteen." *Back row*: **Collins**, Cunningham, Eisele, Freeman, **Gordon**, Schweickart, **Scott**, and Williams. *Front row*: **Aldrin**, **Anders**, Bassett, **Bean**, **Cernan**, and Chaffee.
Credit: NASA

On November 22, 1963, Kennedy was assassinated, and Lyndon Johnson undertook to ensure that JFK's vision for Moon landings would be fulfilled. Indeed, this tragedy only seemed to harden the resolve of everyone to proceed with the rest of the program, which by this time meant the succession of Gemini flights. Gus Grissom had a major part to play in designing the cockpit layout of the two-man Gemini capsule. This craft would have an important role in training crew members who would eventually be going on Moon missions. Gemini made it possible to explore the limits of astronaut endurance to establish if they could remain in space long enough to complete the Moon journey and still remain physically fit. It also made possible the further development of space suits to operate in space, and on the lunar surface, via the exploration of EVAs—otherwise known as "spacewalks." Finally, it became possible, because of Gemini's ability to change orbital parameters, to refine the skills needed for the all-important rendezvous and docking activities that would be key components of the Apollo Moon missions. So, it was natural that, particularly since **Al Shepard** was now grounded, Gus Grissom would fly the first of the manned Gemini missions, known as Gemini 3. His co-pilot for the mission was to be **John Young**, the first of the Group 2 astronauts to "get a ride." Since he eventually made it to the Moon, we need to check out his background.

John Watts Young came from San Francisco but was raised in Florida. His father was a retired Navy Commander. He ended up, as we shall see later, by being perhaps the most experienced astronaut in NASA history, having flown key missions on Gemini, Apollo, and the Space Shuttle. He has a laid-back, "aw shucks," personality and a dry, and even mischievous, sense of humor. **Captain Young** had served in the Korean War on the destroyer *USS Laws*, was a fighter pilot for 4 years flying Cougars and Crusaders, and even set an aviation record (for rate-of-climb) before he became an astronaut. He was particularly noted for precision and attention to detail, and became a safety specialist and advocate among the astronauts. He also was assigned responsibility for pressure suits.[22.6]

Gus had opted to use ejection seats in Gemini instead of the escape tower approach which was used in Mercury, and which was also reinstated for Apollo. Fortunately, none of the Mercury, Gemini, or Apollo missions in the event needed to make use of the emergency exit systems, so we have no objective way of knowing which would have been the more successful. However, we do know that the ejection seats in Gemini took up a lot of space and non-jettisonable mass. However, the seats did bring an air of familiarity to the astronauts who would be spending so many hours in Gemini cockpits, and maybe that allowed them to be able to concentrate on the other matters that they needed to address while they conducted their missions. The "Gus-mobile" was very much designed by a test pilot for test pilots to fly, although **Stafford**, whose height was at the limit for astronaut selection, would later report about the cramped conditions:

"We could not [even] put our feet together in the Gemini feet-well".[22.6]

Figure 2.9 captures well the relationship between the members of this first Gemini team. Grissom, as Commander, had named the spacecraft *Molly*

Figure 2.9. John Young (*left*) and Gus Grissom (*right*), crew of the first Gemini flight, in the *Molly Brown* spacecraft.
Credit: NASA

Figure 2.10. The cramped interior of the two-man Gemini spacecraft. The Commander, by tradition and cockpit design, sat in the left-hand seat; the pilot occupied the right-hand seat. For long-duration flights, storage, sanitation, and cleanliness would be difficult issues to manage.
Credit: Smithsonian NASM

Brown, after the unsinkable character of that name in the Broadway musical comedy, because his previous spacecraft, the Mercury spacecraft *Liberty Bell 7*, had sunk in the Atlantic during recovery operations, nearly costing him his life. Grissom, through this initial Gemini flight, became the first US astronaut to fly into space more than once, but would die during a training accident before he made a third trip. **Young**, however, would eventually make six spaceflights over his long astronaut career, not retiring from NASA until 2004.

We can see for ourselves in Figure 2.10 just how tight was the accommodation for the crew of two in Gemini. This first crewed test flight, known as Gemini 3, also represented the first use of the Titan missile, suitably man-rated, to launch astronauts (Figure 2.11). Astronaut **Stafford** later reported that the Titan launcher:

"produced 8 g loading, and was designed for a nine-megaton weapon."[22.6]

The four orbiting Mercury astronauts had each used a man-rated Atlas booster, so this was to be the first manned test of both the new spacecraft and its launch vehicle.

Figure 2.11. Titan rocket liftoff carrying the first Gemini crew on March 23, 1965. **Young** would later report[25.5] that: *"It really goes up there fast, and MECO [Main Engine Cutoff] staging is pretty exciting."*
Credit: NASA

The Gemini 3 flight gave the all-important thumbs-up to using the Gemini spacecraft as a flight-training test-bed that was going to be needed for all the Apollo techniques intended for meeting Moon mission requirements. The flight lasted almost 5 hours, and demonstrated the satisfactory operation of all major systems and controlled maneuvering, including being the first spacecraft to change plane and size of orbit. The flight also indicated perhaps a slight change in the public's perception of astronauts, which had been created by NASA publicists. During Mercury, with so much catching up to do, compared with the Soviet program, the projected image was of ultra-serious and focused individuals. Now, with the successful completion of Mercury and the initiation of a new phase, perhaps there would be time for a little relaxation, and for the new groups of astronauts to be portrayed as more "human?" **John Young** famously tested that assumption during his first space flight. He smuggled up a clandestine corned beef sandwich in his spacesuit pocket to give to his Commander once they were safely in orbit, something that was in direct contravention of the strict rules about space food being developed at the time! The Mercury crewmen who had been in space long enough to eat at all, would eat only specially prepared, and weighed, meals squeezed out of toothpaste tubes. Times were a-changin'. Space travel was becoming more "routine," and the Gemini crews would bring a different perspective to the endeavor. So what did **Young**, not given to being very verbose, say about his first spaceflight?

"It was really something—twenty minutes to Africa."[25.6]

After successful recovery of the capsule, the by-now customary parades took place, but this time celebrating two crewmen. Grissom later captured the experience in his book *Gemini*,[25.2] which would however end up being published posthumously, as events would develop. Figure 2.12 exhibits part of the asso-

Figure 2.12. A traditional lapel pin records **John Young**'s first spaceflight.
Credit: Author's collection

Figure 2.13. Group 4—"The Scientists," June 1965. Michel, Garriott, **Schmitt**, Gibson, and Kerwin.

Credit: NASA

ciated paraphernalia. The nation realized that America was catching up in the Space Race with the Soviets, and that by using Gemini, step by step, the US was going to indeed be able to make the Moon journey. But would they be first? That was still in doubt, but NASA mission planners had a series of test flights lined up to test all the necessary elements of the eventual Apollo Moon mission. This even affected the thinking regarding the skill sets needed for the astronauts on lunar missions, and therefore the appropriate recruitment criteria.

A fourth group of astronauts, "The Scientists," was selected in June 1965 (Figure 2.13). Only one of the five, **Schmitt**, would eventually make it to the Moon, however. They would have to wait a long time before flying at all. In fact, one of them, Michel, gave up waiting and left NASA before ever being assigned a flight. They were made of different metal than all the previous astronaut groups, all of whom had consisted of military fighter pilots. Some of these new guys could not even fly! It is not difficult to imagine the reaction to this group of newcomers amongst the already established test pilot astronauts.

Now, with two of our eventual Moon travelers having had their first space flights, we're really getting started. The next phase of the space program would involve an intense period of flights, each one further developing the necessary technologies and experience by building upon earlier missions. The public would be able to witness, in a succession of missions, an amazing mixture of achievements, and great imagery, which would, over the short period of only a year and a half, demonstrate that the US had indeed caught up with the Soviets, and had a chance at winning the race to the Moon. The television news channels, radio broadcasts, newspapers, and magazines were eager to report it all in this pre-Internet era.

CHAPTER 3

Intense Training

THE NEXT 10 OF THE EVENTUAL MOON TRAVELERS CHECK OUT EARTH ORBIT

We now move to the extraordinary period of a year and a half in which the 10 increasingly complex Gemini missions took place. Just think about that. Being around in 1966, you could expect to see a new, exciting, and dangerous feat of space exploration every month or so. The press and public devoured these events; the astronauts competed to outdo each other, notching up new records with each successive mission. Gemini was the essential bridge between Mercury and the Apollo Moon missions. Space developments during this period moved at a fast pace. New techniques were introduced, and new risky steps were taken with each mission, which was fascinating to watch, and yet there was nevertheless a common core to the missions which did not change. To recount them in this book for each flight would be repetitive, so we should just record them once up front.

The common factors included processes and even certain rituals. On the morning of launch day, there was a traditional launch breakfast (usually steak and eggs) shared with other astronauts. The crew would then proceed to the "suiting-up" area, before going via NASA transfer bus to the launch pad. They would ascend the elevator to the top of their Titan rocket, and to the White Room where they would be "installed" in the spacecraft by the specialized crew assigned for the purpose, led by the pad leader Guenther Wendt. Part of the tradition generally involved some launch crew gag-gifts, such as extra-large Styrofoam wrenches, to fix any problems that might emerge during the upcoming flight. Problems were always expected; many of them were life-threatening. Then, everyone but the astronauts would clear the tower for launch. The mission itself would be what changed each time, and we shall discuss these specifics throughout this chapter for each flight. Then we return to the common factors—there would be re-entry, the deployment of the 'chutes, die markers, and other recovery aids and the deployment of paramedics from helicopters to secure the spacecraft with its inflatable rescue collar and ensure the safety of the returning astronauts. The crew was then transported to the downrange aircraft carrier that was supporting the mission (some crews opted to stay in their capsule; others left the spacecraft and were hauled up into the rescue helicopter for that journey). On the carrier, there would be welcoming bands and banners, speeches, and medical checks followed by the cutting of a big cake to be shared among the carrier crew. For the earlier missions there might be a telephone call from the President to welcome the crew back to Earth. On returning to *terra firma*, after the formal engineering debrief sessions, there would be parades. There would subsequently be lots of interviews and headlines in *Life* magazine.

There was a great deal of common material *before* launch day, too. One example of that was the naming of spacecraft and the design of mission patches. The crews figured the patches out themselves, sometimes with help from their wives, and sometimes with some informal "polishing" of a basic design being provided by draftsmen elsewhere in the space program. During the Mercury program, all the spacecraft were named, starting with **Shepard**'s *Freedom Seven*. We have seen that Gus Grissom and **John Young**'s first Gemini craft (numbered GT 3) was named *Molly Brown*. But policy changes prevented the naming of subsequent craft in the Gemini program—something which would change back again as soon as the Apollo flights began. Purists will note that there was a complex numbering system for the launches (including the first crewed Gemini flight being numbered 3); some will insist on using Roman numerals, others will be happy with Arabic numbers. For the purposes of this book, I intend to stick with Arabic—noticing that the astronauts themselves were generally ambivalent to the usage, referring, for example, to either Gemini 10 or Gemini X. If it's good enough for them, it's good enough for me. We even have a situation when Gemini 7 flies before Gemini 6, but so be it. I shall merely stick to the numbering system that was used at the time. Just so long as you don't ask me to *explain* it!

One final matter that needs to be discussed and explained before we begin to track the missions, and note the arrival on the scene of each new astronaut in turn (focusing on those who eventually made Moon missions), is Deke Slayton's training process. He set up a process which he maintained and operated all the way through to the Moon missions, and which generally involved keeping teams together. In general, astronauts would first serve as support for an upcoming launch, and this would mean being a "gopher" for the prime crew. Then, for a later mission, the two new astronaut team members would be assigned as "back-up" to the prime crew for that mission. Being on a "back-up" crew involved a great deal of work, because in principal they could be switched into the prime crew slot if the prime crew was incapacitated in some way. Then finally, the new team would be assigned its own prime mission. Much of the work, of both prime and back-up crews, involved long periods inside various mission simulators. The engineers on the outside would try to test the crew by creating various simulated disaster scenarios, and the crews inside would attempt to recover from the problem and arrive safely. Practice, practice, practice. The two crew members (three, in the case of Apollo) became so familiar with each other's thinking, after so many hours in the simulators, that they could work together very effectively as a team. This was why Slayton preferred to keep the teams together; he would rather replace one entire team with another, than switch out one member to be replaced with somebody else. We shall see later that there could be an exception to this general rule.

That, therefore, is the common ground behind each of the Gemini missions. Despite all the training in advance, it was on the missions themselves that most of the learning took place. Being in a zero-g environment is very difficult to simulate on Earth, and so many of the issues that were encountered during the Gemini missions could not really be tackled until the next mission. It turned out that training in aircraft flying parabolic trajectories to provide a few minutes of zero-g, for instance, was not very effective. Fortunately, there was a high mission cadence, so it did not take long before the next team could try to solve the problem encountered in a previous mission.

In each of the Gemini missions, it transpired that at least one of the crew would subsequently become part of a lunar crew—in all of the missions except one, that is. That exceptional mission, known as Gemini 4, would prove to be very successful, and provide the American public with a fund of exciting imagery and confidence, at a time when they had been yet again beaten by the Soviets to the space spectacular of the first spacewalk (by cosmonaut Leonov flying in Voskhod 2). The crew members of Gemini 4 were its Commander Jim McDivitt and its pilot Ed White, who conducted the first US EVA (extra-vehicular activity, or spacewalk), as recorded by the cover art for a model kit in Figure 3.1. The mission took place between 3 and 7 June 1965. This would

Figure 3.1. A model kit of the Gemini 4 Mission, June 3, 1965, including the first US spacewalk with Ed White outside of the Gemini spacecraft. The black part, up front, would return to Earth, and the rear part, or Service Module containing fuel cells, would breakup on re-entry.
Credit: Revell/author's collection

be the first of the Gemini flights that did not include a hardened Mercury veteran as the Commander. Both of the crew were flying this mission as their first time in space. Neither of them would end up making a trip to the Moon. White would die, with Gus Grissom and Roger Chaffee, in the fire in Apollo 1 during a ground test. McDivitt would continue to develop the space hardware for the Moon missions, but remain in Earth orbit. The last Gemini mission would be in November 1966, just $1\frac{1}{2}$ years later, and in between these two flights, America gained all the spaceflight experience it needed to venture onward to the Moon, and incidentally would by then have ensured that a dozen of the eventual lunar mission astronauts would have been given their first taste of spaceflight.

But we were not yet done with Mercury astronauts. The Commander of the next mission, Gemini 5, was Gordon Cooper, who had flown the most extensive Mercury mission, *Faith Seven*, two years earlier, back in May 1963. His pilot for the mission of Gemini 5 was **Pete Conrad**, one of the new Group 2 astronauts.

Charles (Pete) Conrad was a US Navy man, coming from Philadelphia. His father was not, however, from a military background (although he did fly in balloons during WW1!). Captain **Conrad** flew F4H Phantoms, and came to NASA after being a flight instructor at the Navy's test pilot school at Pax River, Maryland, as well as having been a performance engineer at Miramar Naval Air Station in San Diego. Since joining the NASA astronaut team, he had been part of the launch team for Gemini 3, serving as CAPCOM (the acronym referring to the astronaut based at ground control during the mission who would relay communications between the ground and the astronauts in space). He was described as "colorful" by his fellow crew members. Cooper and **Conrad** had come up with a crew patch for the Gemini 5 mission showing a traditional pioneering covered wagon—the initial version also contained the words "8 days or bust" to make it clear that they were trying to push the endurance boundaries. However, NASA management was concerned that the flight might have to be terminated early, so they insisted on the removal of the motto from the patch. Figures 3.2 and 3.3 both show the Gemini 5 crew. They did achieve their "8 day" target, flying from August 21 to 29, 1965, which in itself already demonstrated that a trip to the Moon and back would be sustainable without significant damage to crew efficiency. Also on the mission, Cooper and **Conrad** were able to conduct some radar ranging tests, to be further developed in subsequent missions to allow for rendezvous and docking. Furthermore, theirs was the first use of the newly invented fuel cells in orbit, something which would be an essential part of the subsequent lunar missions, offering longer life than the earlier batteries.

Although he had now made it into space, **Conrad** was not that enthusiastic about the mission, being impatient for more. There had been problems with the fuel cells.[25.2] They were a totally new engineering concept, so this was not unexpected. He later reported:[9.5]

Figure 3.2. The Gemini 5 crew of **Conrad** (*left*) and Cooper (*right*) get onboard for their launch, August 21, 1965.
Credit: NASA

"We weren't busy enough. We weren't that happy with the mission. Most of the experiments had gone by the wayside because of the power-down mode."

NASA management had been right to worry about the length of the mission, which had turned out to be marginal. More work needed to be done, on both men and machines, to test out their capabilities for long endurance. Just to be sure, and to allow for possible problems in the future lunar missions, it was decided to aim for a two-week endurance mission (even though the journey to the Moon and back would only take 6 days).

If **Conrad** was bored on Gemini 5, he certainly would not have been happy on Gemini 7, which would put a new astronaut team to quite a test. They would be aiming for 14 days in orbit, sitting next to each other in the tight space we have seen earlier. Imagine what it would be like sitting with your best friend for even a few hours in such close proximity, let alone having to go to the bathroom in-situ. Two new astronauts from Group 2 would be given this plum assignment—**Borman** and **Lovell** (Figures 3.4 and 3.5). They would prove to be more than equal to the task. They became a great team, both of them eventually making it to the Moon.

Figure 3.3. Conrad (*left*) and Cooper (*right*). Gemini 5 was Cooper's second and last space mission.
Credit: NASA

Frank Borman came from Gary, Indiana, but was raised in the dry desert heat of Tucson, Arizona. He was a no-nonsense Commander, a USAF Colonel, test pilot, and instructor at Edwards Air Force Base in California. He was the astronaut safety specialist, working on the Titan/Gemini escape systems, and also later became the head of the propulsion side of the astronaut office. Prior to getting his mission, **Borman** performed back-up crew and CAPCOM duties on Gemini 4.

Before joining Group 2, **James Arthur Lovell** had been a US Navy Captain, program manager and a test pilot for F4H Phantoms at the Pax River test pilot school. Earlier he had been a safety engineer at a Naval Air Station in Virginia. He was born in Cleveland Ohio, but his father had been killed in a car crash

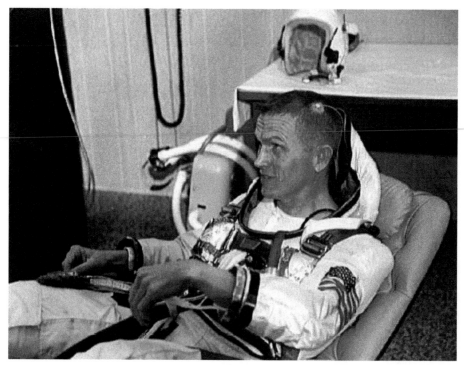

Figure 3.4. Borman gets wired up for the medical experiments before his Gemini 7 launch on December 4, 1965.
Credit: NASA

when he was five, and he subsequently grew up in Milwaukee with his mother. After joining the NASA astronaut team, he carried out back-up crew duties for Gemini 4.

The joint Air Force/Navy crew of Gemini 7 were able to complete their 14-day assignment (Figure 3.6 displays them "counting off the days" while in orbit). They had specially modified spacesuits for their mission with many electrical probes (e.g., Figure 3.4 demonstrates a particularly awkward wiring job on **Borman**). During the 14-day mission, the crew performed 20 medical, technological, and engineering experiments.[6.3] The public began to view spaceflight as routine, and reporting moved away from the former minute-to-minute coverage. **Borman** records:[6.3]

"I looked like a skid row bum recovering from a week-long binge!"

As **Lovell** tells the story,[15.6] once they landed in the ocean, and the recovery crew of paramedics opened the hatch, their rescuers staggered backwards. **Lovell** later asked, following the successful recovery:

"Was this because of our unshaven appearance after our fourteen days in space?"

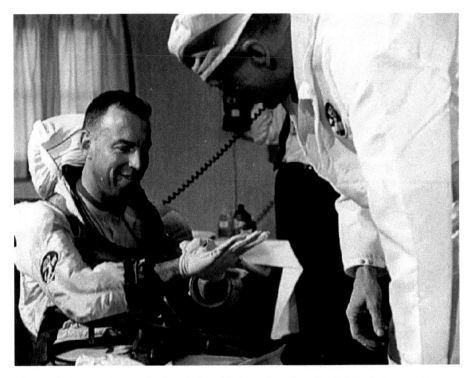

Figure 3.5. **Lovell** gets suited-up before his long-duration Gemini 7 mission, on December 4, 1965.
Credit: NASA

"No sir," the paramedic is reported to have said, "it was the smell!" In addition to the duration aspects of their mission, the crew of Gemini 7 had also been given a new task, that of being a target for a first space rendezvous experiment to be discussed below.

At this point we shall discuss the Gemini 6 mission—which was to eventually fly from December 15 to 16, 1965. Gemini 6 was launched 11 days *after* Gemini 7 due to 3 unscheduled delays. The Commander of Gemini 6 was our old Mercury friend Wally Schirra. It would be quite a mission. Occupying the pilot seat to Schirra's right would be the newest member of Group 2 to get a ride, **Tom Stafford** (Figure 3.7).

Thomas Patten Stafford came from the dust bowl of Weatherford, Oklahoma. He

"wanted to fly since age 5 watching DC 3s flying over, and had my first ride on a Taylorcraft at age 15 or 16."[22.6]

He was a very experienced aviator when he joined the astronaut corps. A Lieutenant General in the USAF, he came to NASA from having been an instructor at Edwards Air Force Base, where he had written many of the test pilot instruc-

Figure 3.6. Hand-written evidence inside the recovered Gemini 7 spacecraft of the crew counting down the days towards completion of their 14-day mission.
Credit: NASA/NASM

tion manuals, following a term flying F-86s in Germany during the Cold War. He was a tall astronaut and there was a need to make a minor modification to the Gemini spacecraft—"the Stafford Bump"—to accommodate him.[22.5] **Stafford**, once he arrived on the astronaut team, was first given back-up crew duties on Gemini 3. Then he was made responsible for range safety, instrumentation, and communications liaison.[22.6]

The mission of Gemini 6 really demonstrated that NASA and its crews could handle flexibility. **Stafford** himself later reported:[22.6]

"We were writing the checklist as we went."

The mission was always intended to test out the ability to rendezvous with a target spacecraft in orbit. But the designated target spacecraft kept changing due to a series of launch failures, first of all the problem was a failure of the Atlas launcher to deliver their Agena target vehicle. Then Schirra and **Stafford** had a launch abort at an amazing 2 seconds from launch due to someone having left a cap on a fuel line.[22.6] The prior experience of Schirra on Mercury was crucial, because a less experienced Commander would probably have

Figure 3.7. Stafford (*foreground*) occupies the pilot seat alongside his Commander Wally Schirra, as they prepare for liftoff of Gemini 6 on December 15, 1965.
Credit: NASA

decided that the crew should eject. They did not; an act of great courage and confidence in the Titan rocket. The crew sat for an hour and a half on top of their fully fueled launcher before they could be extracted. So that then meant that the Gemini/Titan could therefore be recycled for another attempt. Then it was decided to try again and have Gemini 6 rendezvous with Gemini 7, which was already in orbit. Gemini 6 carried the radar, and so **Borman** and **Lovell** were passive targets in Gemini 7. The two Gemini craft were able to perform the first space rendezvous, and then remained in close proximity for 7 hours. The feat was captured in wonderful images and shown in magazines and on postage stamps (Figure 3.8).

The two spacecraft were returned by the *USS Wasp* (Figure 3.9), and the crews celebrated their achievement (Figure 3.10), which was a major accomplishment, making the Moon-landing missions look increasingly possible, because rendezvous was an essential part of the process being developed for the Apollo mission architecture to fulfil Kennedy's challenge. In fact, there had been an intense, even impassioned, debate within the NASA hierarchy about the best way to reach the Moon. Both of the methods under consideration

Figure 3.8. Even inside the states of the former Soviet Union, such as Bulgaria, the joint achievement of the Gemini 6 and 7 rendezvous mission was recognized on their postage stamps.
Credit: Author's collection

Figure 3.9. The two capsules from missions Gemini 6 and 7 are returned to port.
Credit: NASA

Figure 3.10. Reunion of the two Gemini rendezvous crews: Schirra and **Stafford** greet **Borman** and **Lovell** on their return after their long-duration flight. Astronaut chief Deke Slayton looks on, at far right.
Credit: NASA

required the astronauts to be able to repeatedly and safely conduct orbital rendezvous, but one of the approaches needed rendezvous to take place in lunar orbit. We were already 3 years into Kennedy's Moon challenge, but could not make the critical "mode," or "architecture," decision until the astronauts could demonstrate through actual space experience that it was something they could do comfortably, so that even doing it in lunar orbit (as proposed by John Houbolt) was a possibility.

After the success of the joint mission of Gemini 6 and 7, two events which would have an impact on the ongoing program took place. First of all, on February 26, the first test flight of the mighty Saturn 1B rocket was successful. This was the result of hard work from von Braun's team at Huntsville, Alabama. Two of the early Apollo missions which were to remain in Earth orbit would use the Saturn 1B; the later missions which would be headed to the Moon would need the full Saturn 5 rocket, but this was a good start. The second event was a tragedy. On March 2, 1966 two astronauts who were the prime crew for an upcoming flight (See and Bassett) were killed in an air crash while on a visit to checkout their spacecraft. This meant that their mission, Gemini 9, would now have to be flown by its back-up crew, only a month later. With the loss of See, this meant that only one of the Group 2 astronaut contingent still remained to be assigned to a mission. Enter **Neil Armstrong**, who was the last of his group to be selected to a prime crew.

Figure 3.11. **Neil Armstrong** in the cockpit of the X-15 at Edwards Air Force Base in California, before he joined the astronaut corps.
Credit: NASA

Neil Alden Armstrong, coming from Wapakoneta, Ohio, was a civilian when he was recruited into the astronaut corps, being a NASA test pilot flying the X-15 rocket plane seven times out of Edwards Air Force Base, California (Figure 3.11), including once surviving an X-15 landing emergency. He had earlier been a naval aviator conducting carrier landings on the *Essex* during the Korean War—and in fact had ejected from a Panther after flying through an anti-aircraft cable. He once laconically remarked:

"We had a lot of bullet holes in our planes."[4.2]

He was a civilian and his father was an auditor—so he did not come from the usual mold for the astronauts at that time. As a boy, aged 6, he had his first flight in a Ford Tri-Motor. He had performed back-up crew duties on both Gemini 4 and Gemini 5. **Armstrong** was deep down an engineer, with an MS in Aerospace Engineering from USC in Los Angeles. **Frank Borman** described him as: *"intellectually curious."*[6.1] **Armstrong** described the role of a test pilot as:

"Trying to find out where the danger spots are, and make them less dangerous."[4.2]

He obtained his first pilot's license at age 16.[4.2] In addition to living through dangerous episodes in aviation, he had also known tragedy in his personal life, with a child dying in infancy. Although he would become perhaps the most noted astronaut in history, he famously said:

"I would've been reluctant to accept in the middle 1950s that we would see spaceflight in my lifetime."[4.6]

Armstrong was the Commander of the mission assigned to Gemini 8, and his fellow crewman, in the pilot's seat, would be **Dave Scott**, the first of the Group 3 astronauts to get into space.

David Randolph Scott came from San Antonio, Texas, and was an Air Force colonel. His father was an Air Force general—and his fellow astronaut **Cernan** once commented that **Scott**:

"had a military background that goes back forever."[7.3]

He came to the NASA astronaut team from Edwards, as was quite usual for USAF astronaut candidates, and had earlier been a pilot in the Netherlands during the Cold War. He, like **Armstrong**, also had a MS in aeronautics, but in **Scott**'s case it was from MIT.

We do disclose, however, some evidence of a less-serious side to **Armstrong** and **Scott** in Figure 3.12, with perhaps **Conrad** being the instigator. Their mission started well enough on March 16, 1966 (Figure 3.13), but before the day was done the crew of Gemini 8 had to survive a terrifying ordeal which could have been fatal. Only **Armstrong**'s experience in flying X-15s at the edge of space gave him the necessary knowledge base to save their lives.

Gemini 8 was able to achieve rendezvous, with its target Agena, as shown in Figure 3.14. That much had been done before, as we have just seen with Gemini 6 and 7. The next step was to achieve physical docking, and this also they achieved. Docking was described by **Scott** as: *"a real smoothie."*[20.3] Then things started to go badly wrong. The docked spacecraft, slowly at first but then with increasing speed, started to spin. At the time when this happened, their craft was not able to communicate with ground control in Houston. **Armstrong** had to figure out what was causing it, before they lost consciousness. Clearly, **Dave Scott**'s planned EVA had to be scrapped. Initially it seemed that the Agena was causing the problem, so **Armstrong** undocked. However, the spin rate continued to mount—it subsequently turned out that the spinning was due to a thruster which had stuck "open" on the Gemini craft. The spin rate reached 360 degrees per second. **Armstrong** initiated re-entry,

"All we have left is the re-entry system,"[20.3]

he reasoned, and the crew was able to return safely to Earth (Figure 3.15), but it was a near thing. Another astronaut, **Gordon**, later reported:

"Neil activated re-entry without knowledge of Houston."[12.4]

Armstrong later stated that:

Figure 3.12. A moment of goofy comic relief for the Gemini 8 crew of **Scott** and **Armstrong** (*front*), and their back-ups **Gordon** and **Conrad** (*behind*). Wally Schirra, a notorious prankster among the Mercury Seven, had stated that a little levity is appropriate in a dangerous trade.
Credit: NASA

Figure 3.13. A smiling crew of **Scott** and **Armstrong** arrive for their jinxed Gemini 8 mission launch.
Credit: NASA

Figure 3.14. So far so good. Gemini 8 (*foreground*) makes rendezvous with its Agena target, March 16, 1966.
Credit: NASA

"The problem was subsequently fixed for future missions, through engineering."[4.2]

Armstrong's cool head more than once saved his life in emergency situations, and this would be very important to the ultimate success of the mission for the first Moon landing.

So, now the important skills of rendezvous and docking had both been achieved, in addition to all the accumulated experience of long-duration surviv-ability. Deke Slayton, assisted by his Mercury buddy **Alan Shepard**, was still trying to figure out how many crewmen would be needed for the entire astro-naut squad. This was made difficult because of uncertainty in terms of the numbers of future Apollo flights. Slayton decided that there were not enough astronauts under training, and so a new cohort was recruited. They became known, rather ironically, as "The Original Nineteen," (Figure 3.16), and this decision would prove that Slayton's instincts and planning were correct, as half of this new group would indeed make it as Moon travelers (incidentally, **Neil Armstrong** had designed the first crew-assignment schedule for Slayton,[4.2] allowing for the progression from support, through back-up, to prime crew allocations).

Figure 3.15. After a near-fatal emergency in orbit, Gemini 8 with its crew of **Scott** and **Armstrong** are recovered safely. Paramedics surround the craft, and the dye-markers are still producing their green tracers.
Credit: NASA

Figure 3.16. Group 5, "The Original Nineteen" is recruited April, 1966. Nine of the nineteen would become Moon travelers. *Back*: **Swigert**, Pogue, **Evans**, Weitz, **Irwin**, Carr, **Roosa**, **Worden**, **Mattingly**, and Lousma. *Front*: Givens, **Mitchell**, **Duke**, Lind, **Haise**, Engle, Brand, Bull, and McCandless.
Credit: NASA

Of the remaining 10 astronauts in this cohort who did not make it to a Moon mission, five would fly on a Skylab mission, using left-over Saturn hardware to provide for an orbiting space laboratory, (Weitz, Lousma, Carr, Gibson, and Pogue); one would be part of the joint US/USSR ASTP (Apollo–Soyuz Test Project) mission—performing the first hook-up in space between the former Cold War rivals (Brand), two would never fly (Bull left for health reasons, and Givens died off-duty in a car accident); the others (Lind, Engle, and McCandlless) had to wait until the Space Shuttle was flying before they would get into space. In Lind's case, he had to wait 19 years before his Shuttle mission.

The next Gemini flight would be Gemini 9, on June 3–6, 1966, just three months following the recovery of **Scott** and **Armstrong** after the first docking mission. The Commander of this flight would be **Tom Stafford**, who we met earlier with his Gemini 6 mission, accompanying Wally Schirra. This would be **Stafford**'s first mission in the Commander's seat. We have noted the circumstances in which a new crew was assigned to Gemini 9—the March 1966 fatal air crash of the mission's erstwhile prime crew of See and Bassett. **Stafford** and **Cernan** had been the back-up crew, and stepped up into the prime crew slots. They were the first back-up crew to become prime, and fly a mission. There was still a need for more experience of rendezvous, docking, and EVA activities, and this mission was scheduled to provide the opportunity to push the envelope further in obtaining this knowledge. It would be **Cernan**'s first flight, as members of the Group 3 contingent gradually were introduced to spaceflight.

Eugene Andrew Cernan (Figure 3.17), came from modest circumstances in Chicago, his

"Father had to finish High School at night,"[7.3]

and was a captain in the US Navy, flying over 200 carrier landings. He said that:

"Landing at night on a ship—it's just you and your Maker."[7.3]

As we have noted, he was on the back-up crew for this mission before See and Bassett were killed, and had also served as CAPCOM on the Gemini 6/7 missions. He would later describe his experience on this, his first mission, as:

"the spacewalk from Hell."[7.3]

So clearly, things did not go according to plan.

It was planned that the mission would include multiple rendezvous attempts, a docking with an Atlas Agena target vehicle, and a complicated EVA maneuver where **Cernan** would float outside the Gemini, go to the back of the Service Module, put on an Astronaut Maneuvering Unit (AMU)—a kind of wearable jet-pack, and then conduct further experiments in EVA using the unit. The eventual launch is beautifully portrayed in Figure 3.18. However, things went wrong even before the launch.

Prior to that launch, there was a litany of problems, each requiring the crew

Figure 3.17. **Gene Cernan** gets ready for his Gemini 9 mission on June 3, 1966. Maybe he could sense trouble ahead?
Credit: NASA

to get down from the spacecraft, then suited up again and prepared for the next launch attempt. The uncertainty about whether a launch would take place must have been one of the hardest aspects of the astronaut business to handle emotionally. On May 17, their Atlas–Agena target vehicle failed to get into orbit. Then, on June 1, a substitute target was launched successfully, but there were problems with the launch control of their own Titan, and so they missed that launch window. So, they had to get ready again for another shot on June

Figure 3.18. Liftoff of Gemini 9 on its Titan, June 3, 1966.
Credit: NASA

3, and this time their launch was fine. A newspaper headline reported "Gemini 9 Jinx Finally Snapped." However, the newspaper was wrong. Once they got into orbit, they performed the necessary control maneuvers to achieve rendezvous. Figure 3.19 shows what they found on achieving orbit with their target. **Stafford** described it as an *"angry alligator."*[22.10] They considered various possible ways of releasing the target adaptor from its payload fairing which was preventing docking, but the ground controllers decided that would be too dangerous, so the maneuver had to be aborted. Well, there was still the opportunity to carry out the EVA experiments with the AMU to ensure that the mission would provide some benefits. The newspaper headline then read: "**Cernan** set for Sky Walk."

It turned out that even this part of the mission would be problematic, even dangerous. By the end of **Cernan**'s attempts to get into his AMU, conduct his EVA, and return into the Gemini spacecraft, **Stafford** reported later that **Cernan**

"had lost 10 lb."[22.5]

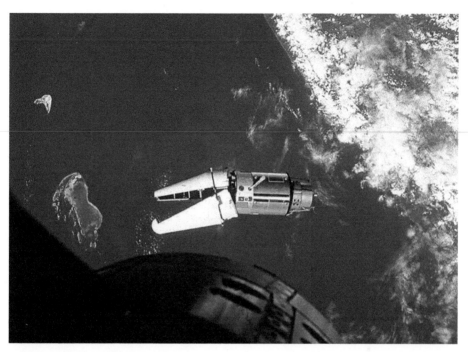

Figure 3.19. "Angry alligator" (payload shroud still attached to rendezvous target) thwarts the docking part of the Gemini 9 mission. Other aspects of the mission involving **Cernan**'s EVA would also prove to be troublesome.
Credit: NASA

The weight loss was due to perspiration as he tumbled about at the end of his umbilical connector trying to gain purchase to do his work. He was never able to get into the AMU, and became physically exhausted. His heart rate monitor registered 170 beats per minute. Also, his helmet visor was totally misted up so that he could not even see as he drifted about. **Stafford** later reflected that

"We did not even think of de-fog in the visor."[22.6]

It was clear that more work was needed to understanding the operational aspects of EVA. **Stafford** had even discussed the issue, before launch with Deke Slayton, of something going wrong and the possibility of having to return to Earth

"dragging a dead astronaut"[22.5]

on the end of his umbilical tether.[22.7] Fortunately, after great difficulty, **Cernan** was able to be squeezed back into his seat, the Gemini hatch closed again, and the spacecraft returned to Earth. Later, **Cernan** asked **Stafford**:

"Would you have let me stay out there?"[22.7]

to which **Stafford** responded:

"Look, Gene, you wouldn't care, you'd be dead!"

In retrospect, this mission seems to have been a total failure so far as achieving its objectives, however at the time the public was just happy to see the new pictures, such as Figure 3.19, and to simply add up the "hours in space" tally that was being accumulated in the competition with the Soviet cosmonauts, in addition to recording another safe return.

Next up to try and tackle the problems of successful EVA operations would be Gemini 10, just a month later on July 18, 1966. **John Young** would get his second Gemini flight, this time flying in the Commander's seat. Flying alongside him, and getting ready to conduct the EVA activities, was **Mike Collins** (Figure 3.20). The aim of the mission was to rendezvous and dock with an Agena, for **Collins** to carry out an EVA and recover an experiment from the Agena and then for the Agena engine to be fired to raise the orbit of the combined

Figure 3.20. Mike Collins (*left*) and John Young get ready for their Gemini 10 launch on July 18, 1966.
Credit: NASA

Gemini/Agena spacecraft assembly. So, how did it work out? It was generally considered successful, as we shall see.

Michael Collins was born in Rome, Italy, and his father was a major general in the Army. Prior to joining the astronaut corps, he was a lieutenant colonel in the USAF, and a flight test officer at Edwards Air Force Base, after having flown F-86s in Iceland and Europe during the Cold War. His first astronaut assignment had been as back-up crew on Gemini 7, and amongst the astronauts he was regarded as the EVA and spacesuit design specialist.[8.6]

Both the crewmen on this mission had a relatively low-key and laconic take on things—they made a good team. They were launched successfully on July 18, 1966, and were able to rendezvous, and subsequently dock, with their Agena target vehicle (Figures 3.21 and 3.22). They were also able to re-fire the

Figure 3.21. The successful rendezvous with the Agena target vehicle on Gemini 10.
Credit: NASA

Figure 3.22. Gemini 10 docks with its target Agena. **Collins** would later recover an experiment from the target vehicle, and the crew would fire the Agena's engine to raise their combined orbit parameters.
Credit: NASA

Agena engine to change their orbital parameters, reaching a new altitude record of 476 miles. **Young** later said:[1.10]

"We were thrown forward in our seats and fire and sparks started coming from the back end of that rascal. The light was something fierce ... I never saw anything like that before ... sparks and fire and smoke and lights."

Collins reported:

"The whole sky turns orange–white, and I am plastered against my shoulder-straps. There is no subtlety to this engine, no gentleness in its approach."[8.4]

The crew undocked from their target Agena, and rendezvoused with a second Agena—the one which had been left in orbit by **Armstrong** and **Scott** when that crew had been forced to abort their mission. **Mike Collins** then, using a jet-gun like Ed White's, rather than **Gene Cernan**'s AMU, for moving around, did a spacewalk to remove an experiment package from the new target vehicle. He had problems with his jet-gun and visibility[8.6] before getting badly tangled up in his umbilical as he struggled to get back onboard his Gemini spacecraft and close its hatch. **John Young** subsequently reported[1.10] that this process:

"Made the snake-house at the zoo look like a Sunday school picnic."

During the maneuvers, **Collins** had managed to lose his camera, which floated away, so there are no photos of Gemini as seen by **Collins**.[8.4] His verbal account is nevertheless illuminating:

"My first impression is a feeling of awe at the wide visual field, a sense of release after the narrow restrictions of the tiny Gemini window. My God, the stars are everywhere. They are bright and steady. Of course, I know that a star's twinkle is created by the atmosphere, and I have seen twinkle-less stars before in a planetarium, but this is different; this is no simulation, this is the best view of the universe that a human has ever had."[8.4]

Clearly, this EVA business was still giving problems. The de-fog, which had been used following **Cernan**'s *"spacewalk from Hell,"* had resulted in **Collins** and **Young** almost being blinded by the chemical once the sun beamed on their helmets. Somehow, all the simulator training in the world was not going to be able to fix this problem. Maybe the next crew, with Gemini 11, would have more success.

Gemini 11 would be launched on September 12, 1966. The mission would repeat the elements of previous missions—rendezvous and docking—which by now were becoming familiar, even including a re-start of the Agena engine to achieve higher orbit. Other aspects would be different. First of all, improved computation methods would in theory make it possible to achieve rendezvous more quickly—indeed within the first orbit, which would give greater confidence for the upcoming Moon missions, provided that the launch could take place within a two-second window. The crew had demonstrated that they could do it in the simulator. Subsequently, the ensuing EVA activity would be extended to make possible tethered flight. This last requirement had, in retrospect, very little to do with the needs of the Apollo architecture, but was an experiment in physics and space engineering which could offer some control solutions for future potential space station applications, as well as perhaps being of use in generating artificial gravity for long-duration missions.

The prime crew for Gemini 11 (Figure 3.23) consisted of **Pete Conrad**, who we have already discussed linked to his earlier Gemini flight with Gordon Cooper, and first-time Gemini pilot **Dick Gordon**. We have briefly observed

Figure 3.23. Gemini 11 Crew of **Gordon** and **Conrad** (*second and third from the left*) discuss the mission around the simulator, with their back-up crew of **Anders** and **Armstrong** (*left and right of group*).
Credit: NASA

Gordon, goofing around with **Conrad** (Figure 3.12), during the Gemini 8 mission preparations, where **Conrad** and **Gordon** were the back-up crew supporting mission Commander **Neil Armstrong** on that flight. Now, for Gemini 11, **Pete Conrad** became the mission Commander, with **Neil Armstrong** as his back-up. In the pilot's seat was **Dick Gordon**. Together they would achieve a new altitude record of 850 miles, and in so doing obtain some spectacular images (Figure 3.24) with which to feed an eager public. Not a bad gig for **Dick Gordon**'s first space-flight. From this altitude, the Earth was really beginning to show its curved horizon, and beginning to look like a sphere. At the time, these were phenomenal images.

  Richard Francis Gordon came from Seattle, Washington, and was, like **Conrad**, a captain in the US Navy. Before he was selected as an astronaut in October 1963, he had been Flight Safety Officer at both the Pax River and Miramar (San Diego) Naval Air Stations. **Al Shepard** had been his instructor at Pax River.[12.3,12.4] He was also an instructor on the F4-H, following action in both the Far East and the Mediterranean. He even held the Bendix Trophy for the transcontinental speed record (869.74 mph) at the time he applied to NASA.[12.4] After joining the astronaut corps, he had performed as back-up on Gemini 8 and CAPCOM on Gemini 9. **Gordon** was earthy, with a dry sense of humor. A typically pointed **Gordon** remark was:

Figure 3.24. **Conrad** and **Gordon** in Gemini 11 use the docked Agena's engine to raise their orbit to an apogee above the tip of India. Images like this were amazing at the time, and eagerly awaited by the readership of magazines, as well as television viewers.
Credit: NASA

"Pete hummed when he was scared!"[12.4]

Gordon has the distinction of being the first, and maybe only, astronaut to fall asleep during an EVA.[12.3]

The Gemini 11 mission called for **Gordon** to perform an EVA and link the Gemini to the Agena by means of a tether. The combined unit of Gemini and Agena joined by the cord, would then be able to carry out experiments in gravity gradient stabilization. **Gordon** later reported that:

"the spinning approach worked at creating artificial gravity."[12.4]

Gordon, despite having the familiar visor problems,[12.3] was able to scramble from the Gemini across to the Agena, grip it between his legs, giving rise to a

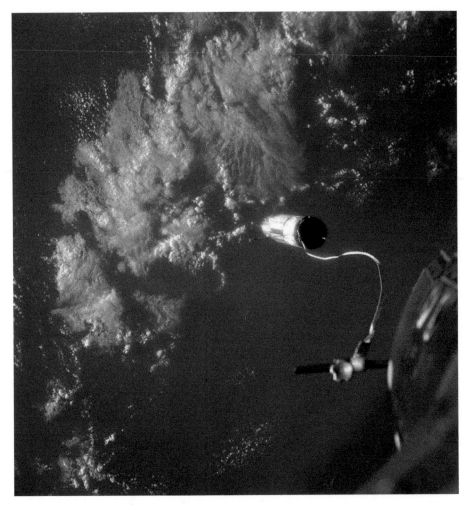

Figure 3.25. Gemini 11 makes tethered flight during its mission on September 12–15, 1966.
Credit: NASA

"Ride'em cowboy!" remark by **Conrad**,[12.3] and make the connection between the two craft (Figure 3.25). So, the Gemini program was delivering the goods, and **Gordon** was the latest astronaut to gain space experience from its missions. He afterwards remarked that:

"EVA feedback from previous missions was poor,"

and that he

"did not know of the problems his predecessors had had until books were written."[12.4]

One assumes that this was at least in part due to the "jet jock" mindset amongst all astronauts, who did not wish to admit that they were anything less than perfect specimens of test pilots. After the recovery of the spacecraft, **Gordon** joined **Armstrong** at the request of President Johnson on a goodwill tour of South America. This was just one of those extraneous duties that astronauts were sometimes selected to perform in that era, between flight assignments.

We move now to the last Gemini mission, Gemini 12, which was to fly November 11–15, 1966. This was just three years from the deadline the President had set for achieving a lunar landing. Nobody was talking about relaxing that deadline, which held firm, something which underlines the continuing all-pervading Cold War mentality. By now, much experience had been gained in the separate skills of rendezvous, docking, and EVAs, and it was hoped that this final mission would be a convincing confirmation that lessons had been learned. Indeed, this would prove to be the case. For the crew, **Jim Lovell** would become Commander, after his Gemini long-duration mission with **Borman**. For his pilot, occupying the right-hand seat and performing the EVA's, he had a very different astronaut from within the Group 3 astronaut cadre, **Buzz Aldrin** (Figure 3.26). As back-up crew he had **Cernan** and Cooper.

Figure 3.26. Gemini 12 prime crew of **Aldrin** and **Lovell** (*front row*), with their back-ups **Cernan** and Cooper.
Credit: NASA

Cooper would subsequently become back-up Commander for Apollo 10, but in fact never flew in space again. We had nearly seen the last of the Mercury astronauts by now, apart from of course **Shepard**, who would eventually fly again.

Edwin Eugene Aldrin (he later changed his name legally to **"Buzz,"** which had been his nickname), hailed from Montclair, New Jersey. His father was a former colonel in the Army Air Corps. He was a USAF colonel, and had flown 66 combat missions with F-86s in Korea, downing two Mig aircraft. Prior to joining the astronaut group, he had also been aide to the Dean of Faculty at the USAF Academy. Perhaps of most relevance to his contributions within the Gemini, and indeed subsequent Apollo, program were his skills in calculating rendezvous maneuvers. **Buzz** was a Doctor of Science from MIT, and the title of his thesis was: "Guidance for Manned Orbital Rendezvous." They called him "Dr. Rendezvous." He was rarely seen without his slide rule (Figure 3.27). He had served as CAPCOM on Gemini 5, then as CAPCOM and back-up crew member on both Gemini 9 and 10.

One of the innovations that **Aldrin** brought to the program was buoyancy training, as a way of handling the effects of zero-g on Earth. Astronauts would spend time submerged in a huge tank of water which contained mockups of the spacecraft hardware elements as they would be configured in orbit. This meant that astronauts could attempt to carry out steps they were to use in orbit. They soon realized that many more hand-holds were needed outside the Gemini and on target spacecraft. With these additions, the EVA became a relative breeze. During Gemini 12, **Lovell** and **Aldrin** were able to conduct three dockings and extensive EVAs with practiced success (Figure 3.28). Furthermore, **Aldrin** demonstrated that they could perform rendezvous solely by hand calculation, which would be important in the event of equipment failure. In addition, one more Gemini tether test was conducted. At one point, **Aldrin** wiped clean **Lovell**'s window, and **Lovell** said:

"Hey, would you change the oil, too?"[1.10]

It had finally been proven possible to demonstrate that an EVA could be successfully conducted, with appropriate training and techniques. **Aldrin** said:

"I think some of the advantages offered by the underwater training facility that I was involved in gave me a considerable advantage."[2.3]

With this mission, the Gemini program was able to end on an upbeat (Figure 3.29).

The Gemini missions are summarized in Appendix C. Gemini had therefore shown that a rendezvous in orbit was not intrinsically a difficult task, and so a final decision was made on the Moon mission architecture—it would use the lunar orbit rendezvous approach, one which was suggested by Houbolt as being the more energy efficient architecture.

By now, half of the 24 astronauts who would eventually make it to a lunar mission, had their baptism of spaceflight, and some of them had flown more

Figure 3.27. This painting of **Buzz Aldrin** during the Gemini 12 mission, with his ever-present slide rule floating in front of him in zero-g, is the work of space artist Pamela Lee, who calls the piece "Gravity."
Credit: Pamela Lee

than once. A strong and experienced cadre had been assembled and trained, and they were all jostling for an opportunity to be assigned one of the Moon missions. Indeed, at this time, it was unclear how many Moon missions there would be. As we pointed out, President Kennedy's injunction would be satisfied

Figure 3.28. Aldrin finally overcame the problems of EVAs that had been experienced throughout the Gemini missions. On Gemini 12, he was able to accomplish all the tasks during extensive EVAs, thanks to better preparation, training, and design modifications.
Credit: NASA

with *only one* successful Moon landing and return to Earth. These missions would be very expensive, but there needed to be enough hardware (spacecraft and rocket stages) in the pipeline so that by steady progress, a Moon landing would eventually be achieved. Meanwhile, the astronauts were relying on Deke Slayton's decision-making process to ensure that they would be assigned either as a back-up, or ideally as a prime crew member, for the upcoming Apollo missions. At this point, several of the Group 3 astronauts, and all of the Group 4 and 5 astronauts had still to fly. The Apollo missions, at least, would each involve 3 crewmen, so that would open up opportunities. Nevertheless, the unavoidable implication is that the last 12 of the 24 astronauts who would

Figure 3.29. **Aldrin** and **Lovell** bring the Gemini program to a successful end on November 15, 1966, when they arrive on the deck of their recovery carrier, the *USS Wasp*, after Gemini 12's successful re-entry and splashdown.
Credit: NASA

eventually fly to the Moon, would have to do so *on their first flight*! They would be rookies to the Moon. Figure 3.30 is the postage stamp—actually a pair of stamps—which was issued to celebrate the success of the Gemini program.

However, all of this intense action conducted throughout the Gemini program was about to come to a grinding halt. In January 1967, during a ground test of the new Apollo spacecraft, a fire broke out and the astronauts were trapped inside in the oxygen-rich atmosphere with a hatch that could not be opened quickly enough to save their lives. Veterans Gus Grissom, Ed White, and new astronaut Roger Chaffee died in the fire. Gus had said just over a year earlier: "If we die, we want people to accept it. We are in a risky business, and we hope that if anything happens to us, it will not delay the program. The conquest of space is worth the risk of life." There would be an investigation. It would be two years later, in October 1968, before the space program would recover and astronauts would fly again. Meanwhile, there was a complete re-think of the way the Apollo spacecraft had been designed. This two-year

Figure 3.30. A pair of US postage stamps celebrate the Gemini program in 1967. Designed by artist Paul Calle, who managed to create stamps which work either as a pair or when used separately. Gemini had been a very successful program providing astronauts and engineers with key experience and training that was needed for the subsequent Apollo program.
Credit: Author's collection

delay put at risk the probability of America being able to win the space race by landing an astronaut on the Moon and returning him safely to Earth before the decade was out, and before the Soviets. At the time, it was still not at all clear where the Soviets were in their own efforts. It turned out after the event, that the Soviets were having their own problems, including having massive explosions of new vehicles on the launch pad. It was a time for national mourning, and for reflection. *Life* magazine covered the funeral, and then paused for a year, before picking up the story again with Wally Schirra and Apollo 7.

In the midst of all this re-assessment of hardware for the Moon missions, another group of astronauts were recruited in August 1967, but this would turn out to be a poor call by Deke Slayton. They were the Group 6 astronauts, unfortunately subsequently named "The Excess Eleven!" *None* of them would fly to the Moon. In fact, only four of them would get into space at all, after a wait of 15 years from their date of selection.

So, to get to the Moon, we are going to have to use the guys who are already on board. We shall see how they did it in Chapter 4.

CHAPTER 4

The Moon Missions

THE REMAINING 12 MOON TRAVELERS ARE ALL ROOKIES TO THE MOON

This chapter starts in the aftermath of the Apollo fire, which took the lives of the Apollo 1 crew (Figure 4.1). The fire happened in January 1967, and the next flight did not take place until October 1968, so much ground had been lost in the race toward completing President Kennedy's challenging timescale. Several of the astronauts were involved in trying to determine the reasons for the fire, and its implications in redesign. In particular **Frank Borman** took a leading role in the investigation. A new hatch had to be redesigned, which could be more easily opened in an emergency (and indeed whenever any future flight would require an EVA (extra-vehicular activity)). The interior of the Apollo command module would need to have its surfaces replaced with flame-retardant materials. Better quality controls were introduced throughout the engineering and manufacturing processes. So, when flights recommenced in October 1968, astronauts effectively had a totally new spacecraft. The Commander of this re-instated mission for Apollo would be the highly experienced Wally Schirra, who was determined to complete the job that his former Mercury buddy Gus Grissom had started, in proving the spacecraft in orbit. If successful in October 1968, then there would only be just over a year remaining in order to carry out the first Moon landing. It would turn out that there would eventually be 11 Apollo flights, with 9 of them heading off to the Moon, with the last being in December 1972.This chapter will cover a period of only four years, during which a great deal was achieved.

So, we start the final phase of the Moon challenge with the first crewed mission after the Apollo 1 fire, which was named Apollo 7, and was one of the two Apollo missions which remained in Earth orbit. Apollos 2–6 were uncrewed test flights. Apollo 7 was the first US three-man mission, it took off on October 11, 1968 (Figure 4.2), lasted 11 days, and demonstrated that all of the Apollo Command Module spacecraft systems were functioning satisfactorily. Perhaps, surprisingly, to a modern reader, this was also the first mission to have live TV from inside the spacecraft, which was significant enough at the time to merit front-page newspaper headlines. It thereby put the program back on track after the Apollo 1 accident and the resulting almost 2-year hiatus for

Figure 4.1. Arlington National Cemetery, Washington DC. Fellow military officers and astronauts **Shepard**, Glenn, Cooper, and **Young** attend the caisson carrying Gus Grissom's casket following his death in the Apollo 1 pad fire of January 27, 1967.
Credit: NASA

review and retrenchment. The Mission Commander Wally Schirra had picked up the baton from his former buddy Gus Grissom, and completed the test flying program which Gus had started. In so doing, Wally had become the one and only Mercury Seven astronaut to fly on Mercury, Gemini, and Apollo. At this time, Gordo Cooper was still in the running to replicate that feat, acting as back-up crew Commander for the upcoming Apollo 10 flight. But it transpired that Gordo would not fly again, so the accolade goes to Schirra alone. The full Apollo system comprised both the Command Module (with its Service Bay) known as the CSM, and the Lunar Module, or LEM. As a matter of nomenclature, the terms Lunar Module, LM, and LEM were used equally in text and speech at the time. In this text, both LEM and LM are used interchangeably. Formally, the relevant crew members were designated as LMP for the Lunar Module Pilot, and as CMP for Command Module Pilot. The two separate spacecraft—the CSM and LEM—needed to be able to separately operate and rendezvous and dock with each other, which is why there was so much focus on rendezvous and docking during all the Gemini training missions.

Figure 4.2. Liftoff of the first crewed Saturn, a Saturn 1B, carrying the Apollo 7 crew of Schirra, Eisele, and Cunningham, into Earth orbit on October 11, 1968.
Credit: NASA

The Apollo 7 mission did not fly the full Saturn 5 stack, since it was not carrying a LEM, nor headed out of Earth orbit. It did nevertheless provide a good test of much of the Saturn launch vehicle hardware and pad infrastructure, including the new Vehicle Assembly Building (VAB) and the massive

Figure 4.3. View of the Cape launching pad looking down from the Apollo 7 Command Module in orbit. The upper stage of the Saturn is seen following along behind, carrying a LEM simulator (seen inside the opened petals) for testing of docking maneuvers between the Apollo Command Module and the eventual LEM, which was as yet unready for spaceflight.
Credit: NASA

crawler vehicles that transported the erected launch vehicle from the VAB to its launch pad. One of the outcomes of the mission was the decision to have the panels at the top of the Saturn upper stage totally separate and detachable to reveal the Lunar Lander before docking could be safely attempted. In Figure 4.3 we can clearly see that it was not sufficient to have them merely open like flower petals. During the mission, Schirra famously developed a cold, and ground controllers for the first time had to wrestle with the appropriate procedures under the circumstances of a zero-g environment. Schirra, in the event, decided he would not wear his helmet on descent, in case a sneeze would have

Figure 4.4. The crews (*front row*) of Apollo 7 (Cunningham, Eisele, and Schirra) and Apollo 8 (**Anders**, **Lovell**, and **Borman**) in the White House to meet with Charles Lindbergh (*back left*) on December 9, 1968. Also in the photo are President and Mrs. Johnson, NASA Administrator Webb, and Vice President Humphrey. The photo opportunity took place after Apollo 7 had flown, and just before Apollo 8 headed off for man's first journey from the Earth to the Moon, a mere 42 years since Lindbergh's historic first crossing of the Atlantic.
Credit: Lindbergh Picture Collection, Manuscripts and Archives, Yale University Library

been suffocating. This would be Schirra's last hurrah, as he had decided even before launch. He would not fly again. Neither would either of his fellow rookie crew members Cunningham and Eisele. We record them in Figure 4.4, two months later, at a special event at the White House. Charles Lindbergh came to meet the crew of Apollo 7, and the crew of the upcoming flight Apollo 8, to provide a very human link between the first era of aviation and the missions to the Moon. Lindbergh flew solo from New York to Paris in 1927, just 21 years after the Wright brothers made the first controlled powered flight.

President Johnson had taken over the responsibility to keep the Apollo program funded, and to keep on track the Moon landing, since he assumed the presidency after Kennedy's assassination. He would ultimately witness the success of the Apollo 11 Moon landing, though no longer as President. It would be President Nixon whose name would be on the plaques attached to the Moon Landing craft.

Meanwhile, in that Cold War era, it was difficult to know what the Soviets were doing. They did not conduct as many flights as did the US during its Gemini program. However, they were still capable of surprises, and had flown a 3-man craft. There was some indication from intelligence sources about sus-pected Russian plans which would have a profound effect on the developing Apollo program. It changed the character and intent of Apollo 8, turning it into the dramatic mission which Lindbergh was commemorating in Figure 4.4. For the first time, humans would leave Earth's gravitational field and head out to the Moon, where they would orbit, and then return. There was some think-ing at the time that such a feat would almost "count" as the US having done what Kennedy asked, just in case the Soviets pulled off a landing. This was probably the most risky of all the missions. It required the first flight of the full Saturn 5 booster, in "all-up" mode, which meant instead of incrementally testing each stage with its own test flight, it would be an "all or nothing" launch—to save time and money, the previous approach of step-by-step testing of rocket stages was circumscribed, and the full Saturn 5 was used to send the Apollo 8 crew to the Moon (Figure 4.5). The crew found the rocket ride memorable:

Figure 4.5. Launch of Apollo 8 on the first crewed Saturn 5, December 21, 1968, carrying the first humans to visit the Moon—**Borman**, **Lovell**, and **Anders**. All 24 of the eventual Moon travelers would make their journeys in the tiny conical Apollo capsule sitting on top of this massive vehicle, and that capsule would be the only part that would return to a safe landing on Earth after the adventure.
Credit: NASA

"I felt like a rat in the jaws of a giant terrier. The noise was unbelievable," said newbie crew member **Anders**.[3.5]

Once in orbit, and after firing the motor for the trans-lunar injection (TLI), **Lovell** recorded:

"There wasn't much sensation of what was really happening, except that we'd look at the computer and it was adding up the velocity. But it really hit us when we turned around and looked at the Earth. From the time it took us from the burn until we got around to see the Earth again, it had started to shrink. We were on our way to the Moon."[15.5]

The crew consisted of the tried-and-tested team of **Borman** and **Lovell**, together with a new "rookie" astronaut **Bill Anders** (Figure 4.6). Somewhat

Figure 4.6. Bill Anders is ready for his Apollo 8 mission, his only spaceflight.
Credit: NASA

ironically, on this his only flight into space, **Anders** was the LMP, on a mission which did not carry a LEM.[3.1] He had of course been training for the task of LMP, including flying the temperamental Lunar Landing Research Vehicle (LLRV), for some time since he joined the program in October 1963. However, the high-level decision to change the mission in response to perceived Soviet plans meant that his proposed LEM would not be ready in time for the flight. The Apollo LEM had still not even flown in Earth orbit at the time of the Apollo 8 mission.

William Alison Anders was born in Hong Kong, during the British Mandate, because of a posting by his father, who was a USN Commander. His father was subsequently stationed in Bremerton, Washington State, and this became the favored locale for **Anders** later in his life. **Anders** held several flight records and was a Major in the USAF. He flew F-89s over Iceland during the Cold War, and subsequently, perhaps ironically, spent time in Iceland doing geology field work as part of the training program put in place by Deke Slayton for all potential lunar astronauts, as the Apollo program continued. He gained an MSc in nuclear engineering, and was an instructor pilot at a USAF weapons lab in New Mexico at the time of joining the astronaut program. His first back-up crew assignment was for Gemini 11, as we have already noted in Figure 3.23, and he also served as CAPCOM on Gemini 4 and Gemini 12. Regarding Gemini 11, he had said:

"The bad news for me was that Lovell and Aldrin did such a good job on their Gemini flight that they canceled the one that Neil [Armstrong] and I were going to fly."[3.5]

Currently, **Anders** has several passions, including fishing and flying. He owns a de Havilland Beaver, and competes in Reno air races in a P-51 Mustang he names *"Val-Halla"* after his wife, Valerie, and his military code sign "Viking"(-Figure 4.7). He now runs the Heritage Flight Museum with his son, Greg.

Apollo 8 would prove to be a major success. They were the first crew to receive the TLI instruction:

"OK, Apollo 8, You're go for TLI"

the jargon term for Trans Lunar Injection. In this case the instruction came from CAPCOM **Mike Collins**, who was originally on the mission, but was removed because of a bone spur issue (subsequently resolved with neck surgery). The crew orbited the Moon 10 times, in two-hour-long orbits, before returning to Earth to be recovered by the *USS Yorktown*. The views of the lunar surface were spectacular for terrestrial audiences, but only after the crew returned with the film. There was some primitive TV on board, and it was captivating to watch, but most of the public appeared more fascinated by the images of the diminishing Earth. I remember at the time even taking photographs directly from the TV, blurred and black and white though they were, because it was so amazing that humans were so far away. However, it would be the image entitled "Earthrise" (Figure 4.8) which came to exemplify the

Figure 4.7. **Bill Anders'** passion for vintage "Warbird" aircraft—his P-51 *"Val-Halla,"* named after his wife Valerie.
Credit: Brian Lockett

mission, once the film had been returned to Earth and processed. In reality, of course, there is no such thing as "Earthrise" on the Moon. To someone on the Moon's surface, the Earth always hovers at exactly the same spot in the sky above. But, in circling the Moon, the astronauts would see the effect of Earthrise each time they "came around the corner" and saw it again. The photo became iconic, and was used on a US postage stamp to commemorate the Apollo 8 mission. Lovell reported at the time:

"The Earth from here is a grand oasis in the big vastness of space." [6.3]

It was from the crew's spoken words, however, that the Earth-bound observers would learn about the Moon for the first time. **Anders** described the lunar surface:

"like dirty beach sand, pulverized. It looked like a war zone. The back of the Moon is rough as Hell." [3.4]

Borman reported:

"I think the Moon resembled what Earth must've looked like before there was life. Or what it could look like after an all-out nuclear war. So that was sobering." [6.4]

Lovell was particularly impressed by the engineering involved in their mission:

Figure 4.8. The famous "Earthrise" picture, taken from Apollo 8. There was considerable light-hearted banter over the years (and even decades) between the crew members, concerning who amongst them had actually taken the iconic photograph. A journey to the Moon had, because of this image, rather ironically produced a whole new awareness of the fragility of the home planet Earth.
Credit: NASA

"The ground told us that we'd lose radio communication at a certain time, down to the second, and by gosh, they were right!"[15.1]

He later described Apollo 8 (even though he would have one more space mission) as

"the high point of my career."[15.7]

The flight ended 1968 (which had otherwise been a terrible year in the US in-

Figure 4.9. The Apollo 8 crew **Borman**, **Lovell**, and **Anders**, at the Smithsonian's National Air and Space Museum in Washington, DC, on November 13, 2008 (forty years after their mission).
Credit: Author

volving civil unrest and assassinations) on a high note, at Christmas, and **Borman** radioed:

"From the crew of Apollo 8, we close with good night, good luck, a Merry Christmas and God bless you all—all of you on the good Earth."[6.3]

The former crew still meets on occasion to celebrate their mission (Figure 4.9) and to continue their team-based solidarity and ribbing from long ago—**Anders**:

"The famous Earthrise pictures that Jim Lovell keeps claiming he took, but that I was lucky enough to take on our fourth orbit ...."[3.5,15.8]

The US public had been captivated by the first Moon mission—Apollo 8. But that first trip was something of an anomaly, because at that time all the spacecraft elements of a full Moon mission, in particular the Lunar Lander or LM, had not in fact been tested, and the circumlunar trip in some respects was an insertion to reap some public relations rewards while much needed development work continued on the future Lunar Lander. So one more mission would be needed in Earth orbit in order to test out the functioning of the LEM. That would be the job of Apollo 9 (Figure 4.10) in March 1969.

Figure 4.10. At a press conference, the crew of Apollo 9 (Schweikart, McDivitt, and **Scott**) demonstrate models of the Apollo Command and Service Module (on the table), and the two-part Lunar Lander—the LEM—(in **Scott**'s hands), which they would be the first to fly in space.
Credit: NASA

The crew of Apollo 9 would include one more rookie, Rusty Schweikart, who was the LMP, but who would never ultimately make it to the Moon. Jim McDivitt was the Mission Commander and **Dave Scott** would be the CMP for the mission. Both McDivitt and **Scott** had been tried and tested during Gemini. From this mission onwards, because two separate spacecraft would be operating, each craft would be given a name for identification during communications. The mission was successful. First of all, the crew released the LEM from its Saturn upper stage, turned around, and rendezvoused and docked with it. Then Schweikart made his way into the LEM through the connecting tunnel, emerged out onto the LEM porch, and took a photo (Figure 4.11) of the Command Module in docked configuration, with **Dave Scott** emerging from the CSM hatch. McDivitt and Schweikart then separated the LEM *Spider* from the Command Module *Gumdrop*, fired the descent stage engine (which was throttle-able to allow for hovering on the Moon), kicking the LM into a different orbit, eventually separating 113 miles from **Scott** in the Command Module. Then they released the LEMs descent stage (to simulate

Figure 4.11. The Apollo 9 crew in Earth orbit, March 1969, test out the hardware needed for the Moon landing. Astronaut **Dave Scott** is emerging from the hatch of the Command Module *Gumdrop*, which has been docked with the Lunar Lander *Spider*, a small part of which is seen in the foreground. The photo was taken by Schweikart from the porch of the lander. Mission Commander Jim McDivitt remains inside *Spider*—someone had to remain at the controls of the combined complex!
Credit: NASA

leaving it on the Moon on future missions), fired the LEM ascent stage motor, and reconnected with *Gumdrop* some 6 hours later, thus demonstrating that the first team to land on the lunar surface would be able to take-off again, and reconnect with the orbiting CSM. They gave the new element of the Apollo hardware, the LEM, a thorough workout albeit in Earth orbit. This was a very dangerous mission nevertheless. The *Spider* was the first true "spacecraft," in that it could *only* operate in space. When McDivitt and Schweikart departed in *Spider* on their risky mission to test the LEM, they knew that the only way they would get safely back to Earth was to be able to fly the LEM ascent stage to rendezvous and dock back with **Dave Scott** in the Command Module *Gumdrop*, because *Spider* was not designed to be able to survive re-entry through the Earth's atmospheric. In fact, the two-part LEM was a very flimsy craft designed to minimize mass during its allotted mission of getting two men to the Lunar surface and back again up into lunar orbit. After the separation, the spacecraft were indeed able to come back together again.

"Hey Amigo!" said **Scott** to McDivitt in *Spider*, *"I've got you [in the viewfinder]." "Oh Boy,"* said McDivitt, *"Am I ever glad to hear that!"*[20.3]

Although all aspects of the LEM's flight regime had now been tested with the Apollo 9 mission, there was still some concern about the functioning of its engines that would require more work. Nevertheless, this would be the last Earth-orbit constrained Apollo test flight. If any more testing were needed, it would have to take place in proximity to the Moon.

The crew that was assigned to do this further testing was **Stafford, Cernan**, and **Young** (Figure 4.12). We have met them all before, so this was an experienced crew. Politically, in the US, President Johnson had been replaced by President Nixon. His Vice President, Spiro Agnew, kept close tabs on the developing Apollo missions.

Figure 4.12. The Apollo 10 crew (**Cernan, Stafford**, and **Young**) at a May 18, 1969, pre-launch meeting with Vice President Spiro Agnew who is holding a toy of the cartoon character *Snoopy*, after which the mission's Lunar Lander was named.
Credit: NASA

Stafford and **Cernan** had even flown together before. This Apollo 10 mission would be in effect a full dress rehearsal for the lunar landing. It would be the second of the eventual nine Moon journeys. **Stafford** and **Cernan** would fly the LEM in Lunar orbit, and **John Young** would be the CMP, remaining in the vehicle in which all three of them would return to Earth. They named the Command Module *Charlie Brown*, after the cartoon character, and the LEM *Snoopy* after his faithful beagle who always returned home to his master after adventures—names designed to continue the interest already shown by young folks in the mission. **Stafford** records the amazing experience as they head off to the Moon on TLI:

"The white, twisted clouds and the endless shades of blue in the ocean made the hum of the spacecraft systems, the radio chatter, even our breathing, disappear. You have an almost dispassionate platform—remote, Olympian—and yet so moving that you can hardly believe how emotionally attached you are to these rough patterns shifting steadily below."[22.3]

Young put it in his own signature way:

"Seeing Mother Earth shrinking in size behind you tends to make a person a bit nervous, I can tell you!"[25.6]

Figure 4.13 demonstrates for the first time a LEM in the vicinity of the Moon. **Cernan**, as he swept down in *Snoopy* toward the craters and mountains radioed to CAPCOM **Charlie Duke**:

"We're down among 'em Charlie!"[25.6]

This time, at least, an LMP (in this case **Cernan**) could truly justify their title. It was not, however, a fully ready Lunar Lander, and did not carry enough fuel to conduct a full lunar-landing mission. There were still bugs to be found and dealt with. The craft was still too heavy to undertake a full landing mission. Almost, but not quite yet. They did, however, demonstrate color TV direct from space for the first time during the mission. There was a developing realization that the Moon landing would indeed be possible before the end of the year—still 7 months away—and with further effort and by making the camera lighter, it might even be possible to see the landing on live TV. However, that particular transmission, operating from the lunar surface via the Lander, would almost certainly have to be provided in black and white only.

The flight was not uneventful. When in *Snoopy*, **Stafford** and **Cernan**, in the ascent stage, separated from the LM descent stage, fired the ascent stage engine, the spacecraft then went into gyrations. Back on Earth at ground control the physicians, at their consoles in Houston, monitored a dramatic increase in their heartbeats. Fortunately, **Stafford** was able to figure out what had happened (the two crew members had each performed the same computer update function, and therefore one of them had countermanded the instructions of the other).

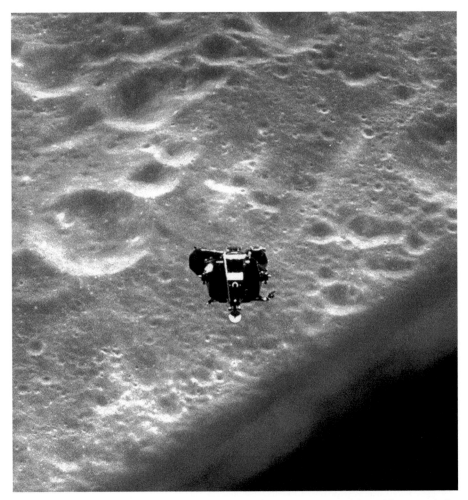

Figure 4.13. The Apollo 10 Lunar Lander *Snoopy*'s ascent stage in orbit around the Moon, carrying astronauts **Stafford** and **Cernan** to within 10 miles of the lunar surface on May 22, 1969. The photo was taken by **John Young**, who remained in the Command Module, *Charlie Brown*, in a 70-mile-high lunar orbit. This image was used for the cover of the June 6 *Life* magazine, under the title "Barnstorming the Moon." They left the LEM descent stage behind, and it eventually crashed into the Moon.
Credit: NASA

"For a while," said **Stafford**, *"things got a little X-rated!"*

After the flight, there would be improved training and changed procedures to prevent the same thing happening again. Many years later, at a talk being given at the Smithsonian's National Air and Space Museum, **Stafford** asserted:

"Now you know why I have no hair—I twice flew with Cernan!"[22.4]

Figure 4.14. The crew of the historic first Moon Landing mission, Apollo 11, **Collins**, **Aldrin**, and **Armstrong**.
Credit: NASA/karsh

Snoopy re-docked with *Charlie Brown*, and the three crew members returned to Earth to be recovered by the *USS Princeton* on May 26, 1969. **Stafford** would not go back to the Moon, although he had another important flight in his future. Both **Young** and **Cernan** had not done with the Moon, however. Each one of them would be going back.

So, after the full dress rehearsal of Apollo 10, the stage was set for Apollo 11. **Armstrong** always referred to it as the first *attempt* to land on the Moon. The crew of Apollo 11 were experienced—each of **Armstrong**, **Aldrin**, and **Collins** having previously been to space (Figure 4.14). They all also had been in dangerous situations. **Collins** nearly went blind while floating outside the spacecraft during his Gemini 10 EVA, and was very lucky to even be on flight status after having had an operation to fix a problem with his neck vertebrae. **Aldrin** had been in dog fights over Korea. **Armstrong** had experienced multiple flying emergencies, including an ejection from his Panther in Korea, an X-15 landing emergency, the Gemini 8 spinning episode, and the Lunar Lander test vehicle crash recorded in Figure 4.15.

They were therefore well suited to this ultimate test of test-flying competence. Former President Johnson was at Kennedy Space Center to see the launch from

Figure 4.15. Neil Armstrong has a near miss when he has to eject (top right) from the Lunar Landing training vehicle (tumbling below) on May 6, 1968—just over a year before his landing on the Moon in the real thing. **Anders** had flown it earlier that day,[3.1] inviting his family along to watch.

Credit: NASA

Pad 39A, thus completing his commitment to his former boss, Kennedy, to see the thing through. **Buzz** described the outward journey:

"We could not immediately detect the fact that the Earth was shrinking as we sped away from it. The Earth would eventually be so small I could blot it out of the universe simply by holding up my thumb."[2.15]

Mike Collins took the photo of what they saw (Figure 4.16). He later described

Figure 4.16. **Mike Collins** took this shot of the *Eagle* Lander ascent stage containing **Armstrong** and **Aldrin** from his vantage point in the Apollo 11 Command Module *Columbia*, July, 1969. Every other human being is on the blue marble in the far distance.
Credit: NASA

it[8.3,8.8] as a photo which showed every other human being in one image—**Neil** and **Buzz** in the Lander *Eagle*, and umpteen billion others back on Earth, and behind him, maybe no other life at all. What an amazing situation to be in. In the actual event,[2.3] there were computer alarms that sounded in the final stages of the landing, and fuel was running low, with the spacecraft headed for a boulder field ... .

Neil Armstrong, of course, pulled it off, landing with only 13 seconds of fuel remaining (Figure 4.17). There are excellent transcripts of the conversations of all the lunar landings.[1.6,1.7,1.11]

This is the conversation (very technical in content, following the checklist, not dramatic or emotional) which took place following those momentous last few seconds of the landing:[1.11]

Aldrin:	*"Contact Light."*
Armstrong:	*"Shutdown."*
Aldrin:	*"OK, Engine stop. ACA out of Detent. Auto. Mode Control, both Auto. Descent Engine Command Override, Off. Engine arm, off. 413 is in."*

Figure 4.17. Finally, the Moon landing took place on July 20, 1969, at Tranquility Base. In this image of one of the Lander's footpads, the contact light probe is seen bent out of shape. **Armstrong** always considered the landing on the surface as the most important part of the mission, with walking on the Moon being a relatively trivial matter in comparison. Credit: NASA

Duke (CAPCOM at Mission Control, Houston): *"We copy you down Eagle."*
Armstrong: *"Engine Arm is off. Houston, Tranquility Base here. The Eagle has landed."*
Duke: *"Roger, Tranquility. We copy you on the ground. You got a bunch of guys about to turn blue. We're breathing again. Thanks a lot!"*

Mike Collins remained orbiting the Moon at an altitude of 70 miles for 13 orbits. He described the view:

"this withered, sun-seared peach pit out of my window. There is no comfort to it; it is too stark and barren; its invitation is monotonous and meant for geologists only"[8.4]

The landing site on the Sea of Tranquility (Figure 4.18) was chosen to be flat, but it did contain some craters and boulders, which required **Armstrong** to take evasive action during the last stages of the descent. It took them $6\frac{1}{2}$ hours after the landing before they were ready to step onto the surface. The whole world was following this, moment by moment. Children were kept awake so that they

Figure 4.18. Armstrong's shadow points across the landing site on the relatively flat Sea of Tranquility to the Lander *Eagle*. **Aldrin** was working on the other side as this photo was taken. This was the farthest either of the crew were to explore beyond their landing site, July 20, 1969.
Credit: NASA

would not miss the historic event. On eventually stepping off the footpad onto the lunar surface, **Armstrong** made his memorable remark:

"That's one small step for a man; a giant leap for mankind."[4.5]

Aldrin best described the view:

"Magnificent Desolation."

Armstrong then reported:

"It has a stark beauty all its own. It's like much of the high desert of the United States. It's different, but it's very pretty out here."[2.11]

The two moonwalkers studied the way that the Moon dust behaved when disturbed. **Armstrong** reported:

"The surface is fine and powdery. I can kick it up loosely with my toe. It does adhere in fine layers, like powdered charcoal, to the sole and sides of my boot."[2.11]

It is hard today to realize just how amazing these events were at the time. Even their fellow astronaut **Alan Bean** reported that he could hardly believe it:

"There's Buzz on the Moon. It was amazing to me because he lived right behind my house!"[5.6]

They deployed six different experiments, collected rock and core samples, erected the flag, and unveiled a plaque:

"We came in peace for all mankind."

They even took a telephone call from the President. They tried different ways of moving about in the $\frac{1}{6}$-g environment. For most of their 2-hour moonwalk, **Neil Armstrong** had the camera clipped to his spacesuit, so when an astronaut appears in the image it is usually **Aldrin**. But, inside their spacesuits and helmets, they looked identical anyway, and **Armstrong** said later that:

"Buzz is more photogenic than me!"[4.7]

On later flights, the Commander's spacesuit carried red markings to make it easier afterwards to figure out who is in any given image. It just did not seem that important to **Armstrong** at the time. He regarded the landing as the key part of the mission, not the floating about on the surface. There is, however, one very good image of **Neil** on the Moon (Figure 4.19). One can imagine his thoughts on reaching this pinnacle of achievement, with nevertheless a number of key steps ahead before a safe landing back on Earth.

Despite concerns just before launch, the Soviet automatic craft Luna 15 did *not* interfere with the mission, and eventually crash-landed on the Moon.[4.11] So the Soviets finished 2nd in the space race. A race where they first seemed far ahead. When it was time to depart, **Buzz Aldrin** transmitted:

"Understand we are number one on the runway."[2.7]

The *Eagle* Lunar Lander Ascent Stage took off, with its payload of two moonwalkers and a trove of moonrocks, and rendezvoused with **Mike Collins** in *Columbia*. They had spent 21 hours on the Moon. Then they headed back on their three-day journey to Earth, and splashed-down in the Pacific being recovered by the *USS Hornet*. In my case, the celebrations on July 24, when they landed back on Earth, were added to those related to my own birthday. I had begun to follow the space program with Sputnik 1 in 1957, when I was a

Figure 4.19. Neil **Armstrong** on the Moon, back in the LEM, after completing his moon-walk EVA. Picture taken by his fellow moonwalker **Buzz Aldrin**, July 20, 1969.
Credit: NASA

twelve-year-old schoolboy. Now, with the completion of the first Moon landing, I was a recently married young man of 24 with a job in the British space program. What a ride it had been! On their return, the astronauts were confined for three weeks in quarantine in case they were contaminated with any Moon organisms. This was an "insurance program" against a very low probability event. The period of confinement and isolation began when they were given biological isolation garments (BIGs) to wear by the paramedics, then continued in a Mobile Quarantine Facility (MQF) trailer on board the recovery carrier *USS Hornet*, and finally in the Lunar Receiving Laboratory (LRL) in Houston. For much of this period of enforced confinement, **Neil** practiced playing his banjo, which seems in retrospect a rather cruel and insensitive (to

Figure 4.20. On return to Earth, the Apollo 11 crew (**Aldrin**, **Collins**, and **Armstrong**) observe one of their returned Moon rock samples at the Smithsonian Institution, Washington DC.
Credit: NASM

his crewmates) way to end the mission. **Armstrong** had taken a piece of the Wright Brothers' first airplane on the mission, and he returned it to the Smithsonian, where it was displayed so that volunteer docents, such as the author, could later show it to visitors, thus linking the two major milestones in the history of flight. Wernher von Braun declared that the magnitude of the event "Could be compared with the time when aquatic life began crawling on land for the first time, and it had ensured mankind's immortality." The rocks were analyzed (Figure 4.20), and told their tale. One of the precious samples went to my old university at Newcastle in the UK, where it was checked out for remanent rock magnetism on a piece of laboratory equipment which I had helped to build when I spent a few months there as a lab technician—so that was my limited contribution to Apollo 11! A new postage stamp honored the achievement (Figure 4.21). The crew undertook a very demanding world tour with their wives, and I was one of those cheering outside the American Embassy in London as they came by on October 14, 1969. By the time the tour was over, everyone was totally exhausted. None of the crew of Apollo 11 ever flew in space again.

Their lives had been changed forever by these momentous events, and we shall note in Chapter 6 what happened to the crew thereafter. One thing became a fact of their lives—they would be forever known as Apollo 11 crew-members, and be required to turn up at various events as living symbols of the

Figure 4.21. Postage stamp issued to commemorate the first Moon landing.
Credit: Author's collection

Figure 4.22. A ritual that was performed every time a new President entered the White House, on the anniversary of the Apollo 11 Moon landing. **Aldrin**, **Collins**, and **Armstrong** meet with President Obama, in the Oval Office of the White House on July 20, 2009, 40 years after the historic event.
Credit: Whitehouse.gov

Apollo era, and that included regular requests to turn up at the White House (Figure 4.22).

The dynamic of the Apollo Program at this point began to change. In a very real sense, the driving purpose of the Kennedy space initiative had been accomplished. The public had been totally absorbed by the process and the achievement, but soon began to lose interest. The motivation on how to continue developed in consultation between the new Nixon administration, NASA leadership, and Capitol Hill reflecting national priorities. There were many Saturn 5 and Apollo spacecraft being completed. A new focus began to emerge involving scientific endeavor. However, the costs—which had amounted to almost 5% of GDP—were not sustainable on an ongoing basis. The Apollo program would have to wind down to more sustainable levels. Eventually Apollo 17 would be deemed to be the last Moon mission. Some spare hardware would be used later for Skylab and the Apollo–Soyuz Test Project (ASTP) missions. Deke and **Al** had a whole squad of astronauts still hoping to get to the Moon. We now continue to record who did eventually make it, what they achieved, and how it affected them. In retrospect, it is quite extraordinary that the program continued after the success of Apollo 11, which had achieved Kennedy's objective, and indeed that there would be six further, risky attempts to land humans on the Moon.

The last of the Group 3 astronauts to receive an assignment was **Alan Bean**, who flew on Apollo 12. He flew with two seasoned veterans, **Pete Conrad** and **Dick Gordon** (Figure 4.23), who had flown together before, during Gemini. In each and all of the subsequent missions beyond Apollo 12 there would be two rookies supporting the Mission Commander. They would all 10 come from Group 5, except for the lone scientist from Group 4, who flew on Apollo 17. The crew of Apollo 12 had been the back-up crew for Apollo 9.

Alan Lavern Bean (Figure 4.24), came from Wheeler, Texas, was a US Navy Captain, and was at the USN Test Pilot School at Pax River when he was selected for the astronaut corps in October, 1963. So he had waited 6 years before his first prime crew assignment. During this period, he had served in back-up crews for Gemini 10 and Apollo 9. He had also served earlier as CAPCOM on Gemini 6/7. **Bean** described himself as a *"plodder."*[5.4] We are very fortunate, in retrospect, that he received this assignment, because after returning to Earth he became an artist who has provided a body of work recording artistically what it was like to be on the Moon.[5.1,5.5,5.7]

Liftoff for Apollo 12 took place on November 14, 1969. So, if Apollo 11 would not have been successful, then this would have been the mission that would still have achieved the "before this decade is out" part of the Kennedy commitment. However, it was President Nixon who was at the Cape for the launch—and it poured with rain. The VIPs were soaked—and when the crew of Apollo 12, in their *Yankee Clipper* spacecraft took off, their craft was struck by lightning (Figure 4.25).

Inside the spacecraft it was dramatic. All the lights went off. The rocket continued to take them higher and faster. **Conrad** had to make a quick decision on whether to abort the mission. Later, fellow astronaut **Charlie Duke** observed:

"It was remarkable that he did not abort."[10.4]

Conrad himself mentioned later:

"After the lightning strike, I was really surprised that they let us go."[9.5]

Figure 4.23. The Apollo 12 Crew of **Conrad**, **Gordon**, and **Bean** with their training aircraft. They had all previously operated as a crew before, as back-up to Apollo 9. **Conrad** and **Gordon** both had spaceflight experience during Gemini 11.
Credit: NASA

Figure 4.24. **Alan Bean** suits up for his Moon mission on Apollo 12.
Credit: NASA

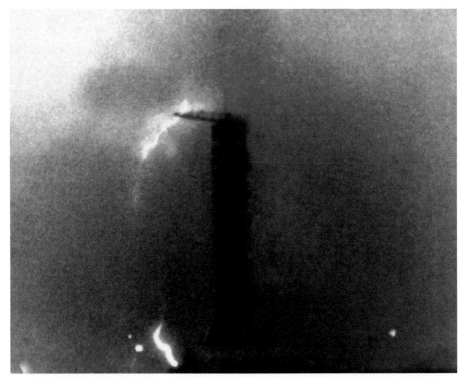

Figure 4.25. Apollo 12's *Yankee Clipper* spacecraft launches off to the Moon, clearing the launch tower in a rainstorm, and is then struck by lightning, which also hit the pad, November 14, 1969.
Credit: NASA

Gordon later drily records:

"We learned not to launch in a thunderstorm."[12.4]

Eventually, with the help of Ground Control, they were able to reset all the systems as they headed off to the Moon. They had never practiced the possibility of lightning strikes in all their previous simulations.[5.6] In lunar orbit, as was by now a matter of routine, the Commander and LMP climbed into the Lander and separated from the Command Module. **Gordon** recounts:

"You know what. I envied them. I wish to hell I could have gone with them, but there was no way for that to happen. I think Pete and Al felt the same way."[12.1]

Bean's own later comments on the subject confirmed this view:

"Imagine training for three years, traveling 250,000 miles to the Moon, then watching your two friends descend the last 60 miles to the surface while you stay behind. Six humans have experienced this frustration."[5.10]

Gordon did say, however, that he was *"happy to be alone"* in lunar orbit.[12.3] The target on the Moon for this mission was the Ocean of Storms, another supposed flat plain. This time, there was an additional element. This was to be a precision landing, made possible by using two deep-space antennas simultaneously providing input to their landing computer, something that was very tricky and risky at the time.[5.6] **Conrad** would attempt to bring the Lander *Intrepid* down in proximity to a *Surveyor* spacecraft which had landed there over two years earlier. **Conrad** made a comparison with a carrier landing:

"You had to get it down or you did not get another chance. I mean, on the aircraft carrier you could always go around."[9.1]

He also noted that, once in lunar orbit:

"There's no doubt in your mind that it [the Moon] was smaller than the Earth. You're only sixty miles above it, and it's all curved away from you."[9.1]

And yes, they did achieve the pin-point landing, as can be confirmed in Figure 4.26.

Then, **Conrad** and **Bean** conducted their spacewalks, which included going to the *Surveyor*, and snipping off pieces, including the on-board camera, to bring back to Earth for later scientific analysis. You can see that camera today at the Smithsonian. They also picked up more Moon rocks (Figure 4.27) and deployed experiments, including a nuclear power generator. On stepping off the lander footpad, **Pete**, who was one of the smallest astronauts, said:

"Whoopie! That might have been a small step for Neil, but it was a mighty big one for me!"[1.11]

Bean joined him, saying:

"OK. My, that sun is bright!"

He later recorded that:

"Running around on the Moon is a lot of fun. It was like you would never get tired."[5.8]

He went on to say:

"The Lunar Module seemed like a house, with Pete and I working and playing all around it. When I think about it now, it was the only 'house' for 239,000 miles."[5.9]

Even though we had the experience of Apollo 11, **Conrad** still later noted:

"I don't think anybody realized what the dust would do up there. I put my foot down and you can just see this great gray cloud go out from it. But it's low. It doesn't billow up, because there's no air. So we got dirty."[9.1]

Figure 4.26. Precision navigation leads to **Pete Conrad**'s pin-point landing, in the Ocea-
nus Procellarum, of Apollo 12's Lander *Intrepid* (on the lunar horizon in the photo) on
November 19, 1969, just 500 feet from the *Surveyor 3* automatic craft which had landed $2\frac{1}{2}$
years earlier. **Alan Bean** is seen inspecting the craft.
Credit: NASA

Bean commented:

*"The craters are hard to see. They look great on the map, but they do not look
worth a damn when you are running along next to them. You can't judge distance,
and you can't tell how far you've run."*[5.8]

Another insight from **Bean** will no doubt cause readers to contemplate how
strange it is to be on the Moon:

*"I have always thought that it was curious that on the Moon all the stars circulate
around you—but once a month instead of once a day here on Earth. Yet the
Earth, which is the biggest object [you see in the sky] there, stays right in the
same spot."*[5.11]

Gordon debated for a while about letting them back into his command module,

Figure 4.27. **Pete Conrad**, in the LRL, eyes some of the Moon rock samples he and **Al Bean** had brought back from the Ocean of Storms.
Credit: NASA

since they were so dirty from their surface activities but relented.[12.3] They then headed off for home, and **Gordon** provides an account of the re-entry into Earth's atmosphere:

"Oh, that's spectacular. You are looking back to where you've been, and that heat shield's burning off, it's a very spectacular sight. It's all kinds of colors. I remember some greens, pinks, yellows. Maybe red."[12.1]

We conclude this account of the Apollo 12 mission looking at Figure 4.28. This is how the future space artist **Alan Bean** remembered seeing the Moon as he headed home. At this point he was still an engineer foremost, though maybe an artist at heart. Listen to how he describes his thoughts on leaving the surface of the Moon to head back to Earth:

"We launch off the Moon. I'm looking out the window. I see us blow these little pieces of foil all around. Gold and silver and black"

Here he is referring to parts of the Lander's flimsy descent stage thermal foil coverage which was ripped off by the ascent stage exhaust. He was not thinking artistically "how pretty," or even "I'm glad the engine is working and I'm getting out of here."

Figure 4.28. Alan Bean's Moon. The Moon as Bean saw it, painting it after he returned to Earth.
Credit: Alan Bean/author's print collection

"What I was thinking was: 'I hope these little pieces of foil don't land on our ALSEP experiment and change the thermal properties'"[5.8]

if they had done so it might very well have wrecked the experiment. No question, **Alan Bean** showed an unusual combination of character traits for an astronaut back then.

The headlines in the newspapers called out "We do it again!" when Apollo 12 landed safely back home and was recovered by the *USS Hornet*. This was beginning to become routine, consequently, the public was rapidly losing interest. All that was about to change.

The story of Apollo 13 was strange in a number of ways. Of course, it made for an excellent book[13.4] and movie with Tom Hanks playing the role of the Apollo 13 Mission Commander, the veteran **Jim Lovell**. **Mattingly** later observed that, although it was a good movie, there were many more compli- cated steps taken to save the crew than could be shown by Hollywood.[16.4] It started off as the mission which **Alan Shepard** had hoped would bring him back to space after he resumed flight status following surgery for Menière's disease (the surgical solution to which had been discovered for him by astronaut **Stafford**).[21.3] He had not flown since his one and only 15-minute spaceflight in 1961, but of course had remained in a position of strong influence within the astronaut's ranks. His Mercury buddy, Deke Slayton, was still in charge of flight crew assignments, and once he knew that **Shepard** was cleared for flight again, he slotted him in to the next available mission, Apollo 13, and provided him with a rookie crew of **Roosa** and **Mitchell**. However, for the first and only time, Slayton was overruled by his management. **Shepard** would need longer to get ready, so he and his crew were moved on to become prime crew for Apollo 14. Lucky for them, as it turned out.

So another crew would become prime for Apollo 13. They had been back-up for Apollo 11. The experienced **Jim Lovell** would lead, with **Ken Mattingly** as CMP and **Fred Haise** as the LMP (Figure 4.29). This mission would be a return to the Moon for **Lovell**, who had already been there during Apollo 8—his two crew members would be rookies.

Thomas Kenneth Mattingly, from Chicago, was a US Navy pilot, who flew from the carriers *Saratoga* and *Franklin Delano Roosevelt* before becoming a student test pilot at Edwards at the time of his selection in April 1966 as a Group 5 astronaut. He had served as back-up, and CAPCOM, for Apollo 8 as well as for Apollo 11. He became a great team player in the Apollo 13 crew. However, with the twists and turns of fate that so impacted Apollo 13, it would turn out that he would *not* fly on that mission. His turn would come later, on Apollo 16. This was because, just a week before launch, the *back-up* crew member **Charlie Duke** (he was back-up LMP) came down with the German Measles.[10.7] Since **Duke** had been in close proximity with the prime crew, the prime crew had to be tested for immunity, and unfortunately **Mattingly** was at risk, so it was decided to replace him. In the event he did not contract the measles, but was very helpful on the ground during the subsequent flight emer- gency. **Mattingly** learned of his substitution from Apollo 13 a week before launch, via his car radio.[16.4]

Fred Wallace Haise, from Biloxi, Mississippi (Figure 4.30), was a former Marine corps fighter pilot who was a civilian test pilot at Edwards at the time of his astronaut selection in April 1966 as part of the Group 5 entry. He had served as back-up crew for Apollo 8, Apollo 9, and Apollo 11, so was very well

Figure 4.29. Original Prime Crew for Apollo 13—**Lovell**, **Mattingly**, and **Haise**.
Credit: NASA

trained. He later said:

"in some ways back-ups were more capable than primes—due to no distractions."[13.2]

He was also CAPCOM on Apollo 8 and Apollo 11. As the LMP of Apollo 13 he would find that he needed all the training he had acquired.

Following the measles scare, **Mattingly** was replaced by **Jack Swigert** at very short notice (Figure 4.31).

John Leonard "Jack" Swigert, from Denver, Colorado (Figure 4.32), had been a fighter pilot with the USAF operating in Japan and Korea, before becoming a commercial test pilot when he was selected into Group 5 in April 1966. He had worked as CAPCOM and back-up on Apollo 7 and Apollo 11, and also on Command Module re-design following the Apollo 1 fire.[24.4] He was selected into the prime crew for Apollo 13 just 3 days before launch. This would be quite a test of Deke Slayton's training, team assignments, and back-up plans, even if nothing untoward was to happen.

Fortunately, for the mission that was to develop, **Swigert** was

"a real good command module pilot,"

Figure 4.30. Fred Haise in the White Room atop the Saturn 5 gantry, getting ready for launch to the Moon on Apollo 13, April 11, 1970. In the background is Pad Leader Guenther Wendt, and **Haise** talks with NASA Suit Tech, Joe Schmitt.
Credit: NASA

as described by fellow astronaut **Duke**.[10.4] It turned out that, after the on-board explosion that would subsequently take place, the crew had to use the Lunar Module in "lifeboat mode" to bring them safely back to Earth proximity, and during all his previous training and simulator work *"Jack wrote the 'lifeboat' manual."*[13.2]

There followed an intense, though short, period of handover (Figure 4.33), before the new crew took off for the Moon on April 11, 1970. **Haise** recounts:

"We were in the simulators to go through all the dynamic mission phases, to make sure we were talking the same language, until about eight o'clock the night before launch."[13.1]

The public had lost interest in the Moon missions, and even though the crew did a telecast from their spacecraft during the TLI phase, the TV networks did not carry the broadcast. The crew switch had taken place so quickly, that **Swigert** had not even had time to send in his tax returns, something that worried him in the early stages of the flight.

"How do I apply for an extension?"

Figure 4.31. New crew for Apollo 13. **Lovell, Swigert** (replacing **Mattingly**), and **Haise**. This new crew did not even have time for the traditional photograph dressed in spacesuits. Credit: NASA

Swigert asked the CAPCOM, as they headed off to the Moon.[23.1] This would become the least of his worries.

After an uneventful trip up to that point—except for a preliminary engine shutdown of one of the Saturn 5 engines following which **Lovell** said:

"Hey, that's our crisis over"[13.2,13.4,15.7,15.9]

everything changed when, halfway to the Moon, the crew experienced an explosion in the Service Module, including a loss of oxygen. There was an array of warning lights and sounds which immediately came on. It was **Swigert** who announced to the ground controllers:

"OK, Houston, we've had a problem here."[1.11]

Figure 4.32. Jack Swigert moves from back-up to prime crew at short notice for Apollo 13.
Credit: NASA

What followed, namely the rescue and safe return of the crew, became known as "NASA's finest hour," and has been recorded well in the Tom Hanks movie. The safe recovery of the crew has been regarded as a

"classic case of crisis management."[13.2]

Figure 4.33. Discussing mission details. Former prime CMP **Mattingly**, and former back-up **Swigert**, before the Apollo 13 mission. **Mattingly** would continue to be intimately involved in the mission, as he worked in the simulators in Houston to figure out ways to bring back the Apollo 13 crew to a safe landing on April 17, 1970, after the explosion in the Service Bay of the Command Module *Odyssey*. The lander *Aquarius* would be used as a life boat to provide life support during their return journey.
Credit: NASA

Fred Haise said later:

"I knew instantly—we couldn't even go into Lunar orbit. So the mission was gone, right there."[13.1]

Ground control worked out procedures for preserving power and oxygen long enough for the crew to swing around the Moon (Figure 4.34) and return to Earth. **Mattingly**, and other available astronaut crewmen, tested out the procedures in the simulators to ensure that they would work. It was a concerted effort with everyone foregoing sleep, working around the clock to make sure of the crew's safe return.

Lovell dramatically recounted later what had happened:

"Suddenly there was a hiss-bang. The spacecraft rocked back and forth, lights were coming on, noise all over, jets were firing. I looked out the window and I could see, escaping from right out of my spacecraft, at a high rate, a gaseous substance."[15.4]

As they eventually swung by the Moon, Lovell was concerned to correctly note

Figure 4.34. The combined Apollo 13 Command Module and Lunar Lander swings by the Moon on its emergency recovery flightpath. **Lovell**, on his second trip to the vicinity, was too busy to look out the window or take pictures.
Credit: NASA

down the new procedures being transmitted from Ground Control in Houston, but his other two crew members were focused on looking at the Moon in close proximity (Figure 4.34).

"They got their pictures; I got the procedures,"

said **Lovell** later.[15.4,15.6] **Lovell** reported:

"A lot of instructions on what to do came from the ground ... It's one thing to set up procedures, but to execute them when you know your ass is on the line is a little bit different."[15.1]

Fred Haise remembers it this way:

"Jack and I had cameras out. We knew that this was the only time we could see it, so we were glued to the windows and shot a lot of pictures. In fact, we shot some of the best pictures in the program, since our flight path allowed us to cover two very prominent seas on the backside—the seas of Moscow and Tsiolkovsky."[13.5]

After a stressful period of powered-down flight—well captured in the Apollo 13 movie—with the entire 3-man crew occupying the 2-man LM *Aquarius* as they returned to Earth, **Swigert** took control of his Command Module *Odyssey* for the re-entry saying to Houston CAPCOM:

"I know all of us here want to thank all you guys for the very fine job you did,"

as they began their uncertain re-entry with a potentially damaged heat shield.[23.1]

Perhaps due to a damaged antenna resulting from the explosion, there was an extended period of communications blackout, adding to the tension while everyone around the world who was following the return of the crew held their collective breath. But all was well.

It is interesting, in retrospect, after their safe return (Figure 4.35), to consider the main concerns of the crew, at a time when most of the world was following the rescue on TV (yes, by now the TV networks had begun to supply broadcasts to an eager public) and just hoping for their safe return. They were returning to Earth in a freezing cold, powered-down spacecraft, with three people in a craft intended to carry two (the LEM). **Haise** had been the intended LMP, descending in the craft to the lunar surface with **Lovell**. Instead it had become a lifeboat. This is what **Haise** later recorded:

"There's no question it was a remarkable recovery from a bad situation. But at the same time, relative to the mission intended, it was a failure. The biggest emotion I had for several months after that flight was disappointment. Just a big sinking feeling ... biggest disappointment of my life. We were very concerned that this flight was now the first failure, and that bothered us, that we may be the cause of the end of the program."[13.1]

The accident investigation revealed that, in the words of **Jim Lovell**:

"The Apollo 13 accident was set up five years before we lifted off ..."

It seems that a tank had been dropped during assembly operations.[15.3] He still supports the continuation of human outreach:

"Spaceflight is so interesting; the results far overshadow the risks."[15.7]

Some of the folklore from that accident and its recovery, however, has had severe consequences in the longer term. The phrase "Failure is not an option," attributed to the ground controller Gene Kranz, was never in fact spoken (as he later confirmed to me). Failure always is an option, and always will be, for

Figure 4.35. Safe return. Apollo 13 crew **Haise**, **Lovell**, and **Swigert** arrive on the recovery carrier *USS Iwo Jima* on April 17, 1970. This would be the last flight for veteran **Lovell**. **Haise** would fly again, in very different circumstances. **Swigert** was done.
Credit: NASA

challenging new endeavors in space. As **Lovell** himself indicated, we need to continue nevertheless. He has, however, earned the somewhat dubious distinction of being the only person who went all the way to the Moon twice, and yet never set foot on it. After the successful return, Grumman, the prime manufacturer of the LEM, sent a mocked-up invoice for $300M to North American, the manufacturer of the CSM, "for services rendered" in bringing the command module home and its crew back alive!

Now, at last, **Shepard** got his chance to return to space, with Apollo 14. After the Apollo 13 accident almost ended the program (after all, the original intent had been achieved with Apollo 11), this would be a mission to restore confidence in ongoing flights to the Moon. **Shepard** as Mission Commander would have two rookies with him: **Stu Roosa** as CMP and **Ed Mitchell** as his fellow moonwalker and LMP.

Stuart Allen Roosa, from Durango, Colorado (Figure 4.36), was a Colonel in the USAF, and a test pilot at Edwards at the time of his recruitment in the April 1966 astronaut selection cohort. He had been a "Smoke Jumper" back in 1953, parachuting into forest fires to deal with them when ground access was

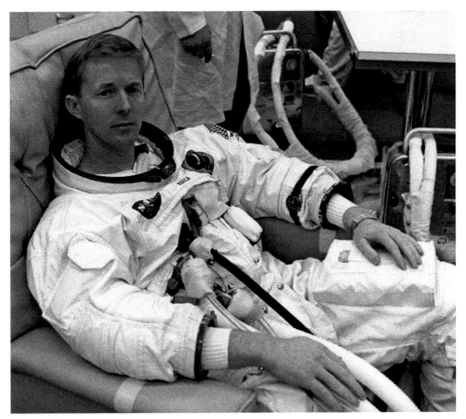

Figure 4.36. Astronaut **Roosa** suiting up for the Apollo 14 Moon flight, on January 31, 1971. **Roosa** was CMP of the *Kitty Hawk*.
Credit: NASA

not possible. He had served in Japan as a Chief of Service Engineering with the USAF. Roosa served on Apollo 9 as back-up, and as CAPCOM on Gemini 12, Apollo 1, Apollo 7, and Apollo 9. He was the

"astronaut specialist on boosters,"

according to his fellow astronaut **Schmitt**.[19.10] Like many of his fellow astronauts, **Roosa** claimed:

"I have wanted to fly airplanes ever since I can remember. I cannot remember when I suddenly decided I wanted to fly airplanes."[18.1]

As CMP, he named his craft *Kitty Hawk*, after the site where the Wright Brothers conducted their early experiments in flight.

Edgar Dean Mitchell (Figure 4.37), was from Hereford, Texas, but spent most of his formative years in Roswell, New Mexico, where his father was a cattle rancher. **Mitchell** was a US Navy Captain, and served on the carriers

Figure 4.37. **Ed Mitchell** in his spacesuit, ready for his Apollo 14 mission to land in the Fra Mauro region of the Moon in the Lunar Lander *Antares*.
Credit: NASA

Bon Homme Richard and *Ticonderoga* in the Pacific.[17.10] At the time of his recruitment into NASA's astronaut corps, he was an instructor in mathematics and navigation at Edwards. His students called him "the Brain." His MIT Doctorate thesis was titled: "Guidance for Low-thrust Interplanetary Vehicles."[17.4] **Mitchell** himself records:

"I spent 9 years collecting credentials before the astronaut selection."[17.10]

Once joining the astronaut team, he became one of the teachers, as reported by his fellow Group 5 astronaut **Worden**.[24.4] **Fred Haise** said he was:

"The leader of Group 5";[13.2]

Mattingly said:

"With his PhD and his style, everybody had a great deal of admiration for him from the beginning."[16.4]

Mitchell worked as back-up on Apollo 9, Apollo 10, and Apollo 13, and he worked on the Lunar Module re-design following the Apollo 1 fire.[24.4] It was **Mitchell** who named the Lander *Antares*, after a star in the constellation Scorpius which was an important navigation milestone during descent to the lunar surface.

Apollo 14 set off for the Moon (Figure 4.38), on January 31, 1971. Because of **Shepard**, who had many high-society friends, there were a number of VIPs at the launch—Secretary of State Kissinger, and actors Kirk Douglas and Charlton Heston among them. It must have been a bitter–sweet moment for Deke Slayton as he watched his old Mercury buddy **Shepard** get his ride to the Moon. **Roosa** describes the outward journey:

"When it really dawned on me that we're a long way from home is when you start picking up the delay in the communications. Then you look out and see this little bitty Earth back here, and you see all that darkness, and you also feel humble at the same time … It's kinda hard to express … that's why we need to send a poet."[18.1]

Mitchell recorded another advance at the time:

"When the public on Earth tuned in, our pictures would be the first full transmission of color TV from the Moon."[17.2] *"We did have a docking problem … but Roosa solved it, although there did persist a nagging worry about the docking mechanism for the return from the Moon."*[17.10]

The destination for Apollo 14 was the same as had been intended for Apollo 13—Fra Mauro, which geologist **Schmitt** described as:

"a very important site on the ejecta blanket of the Imbrium."[19.10]

Mitchell and **Shepard** experienced abort light and landing radar problems on their descent to the lunar surface. None of the Moon flights was completely without tension and emergency. However, eventually they landed.

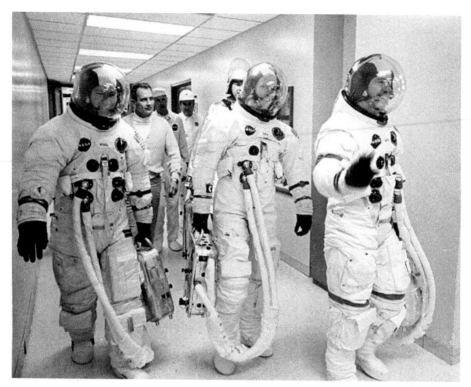

Figure 4.38. The Apollo 14 crew (**Mitchell**, **Roosa**, and **Shepard**) heads out on January 31, 1971, to the launch pad, with Deke Slayton riding herd.
Credit: NASA

"To shut down, and drop to the surface was a real relief,"

said Mitchell.[17.1] **Shepard** records:

"Of course, the first feeling was one of a tremendous sense of accomplishment. A tremendous sense of realizing that, 'Hey, not too long ago I was grounded. Now I'm on the Moon!' "[21.2]

He also later claimed:

"It's a lot harder to land a jet on an aircraft carrier than it is to land a LEM on the Moon."[21.3]

When he stepped onto the surface (Figure 4.39), **Shepard** announced:

"It's been a long way, but we're here!",[21.5]

reflecting on the fact that he had been the first American in space, and would be the only one of his Mercury colleagues to make it to the Moon. He later noted the scenery around his landing site:

Figure 4.39. Al **Shepard** finally makes it to the Moon, stepping off the *Antares* footpad in the Fra Mauro Highlands on February 5, 1971, as Commander of the Apollo 14 mission. **Shepard** looks towards Cone Crater, the destination for the upcoming EVA.
Credit: NASA

"It had a majestic feeling about it—in the sense of a desolate mountain desert type of setting."[21.1]

Shepard, at 47, would be the oldest of the Apollo moonwalkers. **Mitchell** joined him on the surface. His first words, referring to the LEM ladder steps, and maybe lacking somewhat in majesty, were:

"That last one is a long one!"

Roosa, meanwhile, was circling the Moon;

"It looks like a plaster of Paris cast"

was his observation.[18.1] He was very busy:

"I didn't have time to eat. I mean, I had time to eat according to the Flight Plan but didn't. I never fixed one of those freeze-dried meals. I didn't eat very much at all."[18.1]

His comment on being alone in *Kitty Hawk*, while his buddies were on the surface:

"Well, there are only six of us who have orbited the Moon solo, so we've got a smaller fraternity, if you want to count numbers!"[18.1]

Something special that **Roosa** did while on the mission reflects his early experiences of forestry. He took with him a selection of seeds from pines, sycamores, redwood, and Douglas Fir, which afterwards would result in 450 "Moontree" seedlings. More of that later.

On the surface, **Shepard** and **Mitchell** deployed experiments, and collected Moon rocks, and they had a small tool carrier, called the MET, Modular Equipment Transporter, to help them. They sometimes referred to it as a "rickshaw," and sometimes as a "golf cart." Their target was Cone Crater, but they had navigation problems in trying to reach the summit. **Mitchell** reported:

"The clock was ticking, ticking ... and we could not navigate like we thought we could. We never really did know our position within about 5 to 10 meters[17.1] *... Doggone it, you can sure be deceived by slopes here,"*

Mitchell said in frustration.[1.11] They were in fact unable to reach the summit, and had to return to *Antares*. Later, back on Earth, they would realize that they had been just 20 meters from their target, but could not recognize the fairly significantly sized lunar feature, due to the difficulties of lunar navigation. As they prepared to leave, as a moment of lightheartedness, **Shepard** hit a golf ball with a golf club assembled from bits of his equipment:

"You might recognize what I have in my hand as the handle of the contingency sample return; it just so happens to have a genuine six iron at the bottom of it. In my left hand, I have a little white pellet that's familiar to millions of Americans. I'll drop it down. I'm going to try a little sand-trap shot here."[1.7]

He always claimed it went for *"miles and miles and miles."*[21.4] Although, wearing his bulky pressure suit, and only being able to hold the club with one hand, that is unlikely. Some future Moon traveler may find the golf ball and put the question of its distance traveled to rest, once and for all. Then they left the Moon to rendezvous with **Roosa** for the journey home. **Mitchell** noted later:

"Taking off from the Moon, we didn't know what we were going to feel. It was a pretty severe shock. It staggers you. It makes you sag."[17.1]

During the three days of the return flight, **Mitchell**, in the euphoria of the success of the mission, later observed:

"I suddenly had a moment of deep insight. It was an overwhelming realization that my body and my mind were connected to everything in the universe."[17.7]

This is heavy stuff. And **Mitchell** would spend the rest of his life investigating this insight. He also conducted an ESP experiment, but the results were inconclusive. The crew brought back rock samples (Figure 4.40). None of them would fly in space again. **Roosa**'s final comment on the mission was:

"There's probably a sense of pride, more than anything, and you really got that when you arrived on the carrier [USS New Orleans]. We did it! We did it well!"[18.1]

The MET had not been that great a success on rocky surfaces and $\frac{1}{6}$-g conditions. The success of Apollo 14, however, meant that NASA had recovered from the Apollo 13 low point. So the program was going to continue, with the main objective from now on being the pursuit of science on the Moon. An up-rated lander design meant that more payload could be carried to the Moon in the future, and this would include a Lunar Rover, designed by von Braun's team at Huntsville, to enable those on the surface to cover more territory around their lunar base. The first mission to try out these new capabilities was Apollo 15. **Dave Scott** would be the experienced Mission Commander. Included in his crew (Figure 4.41) were two rookies, **Jim Irwin** as LMP, and **Al Worden** with responsibilities for the Command Module, named *Endeavour*.

James Benson Irwin, (Figure 4.42) came from Pittsburgh, Pennsylvania, where his father was a steamfitter. A Colonel in the USAF, he was a test pilot at Edwards when selected to become an astronaut in April 1966. He was deeply religious even before his mission, and that only increased with the experience of his lunar journey. Once on board the astronaut corps, **Irwin** was assigned back-up and support crew duties on Apollo 10 and Apollo 12. He importantly:

"worked on redesign of the Lunar Lander, with Ed Mitchell, at Grumman, Bethpage, following the Apollo 1 fire"

according to **Worden**.[24.4]

Alfred Merrill Worden (Figure 4.43), would be the CMP for Apollo 15. His parents ran a small farm in Jackson, Michigan, and he became a Lieutenant Colonel in the USAF. He flew at Andrews Air Force Base before moving to Edwards, where he became an instructor at that test pilot school. By his own account, he:

"wrote some of the courses. I gave Gene Cernan some training."[24.4]

Worden had also enjoyed some time at the British test pilot school in Farn-

Figure 4.40. Apollo 14 moonwalkers **Mitchell** and **Shepard** in the LRL review their retrieved rock samples from the slopes of Cone Crater.
Credit: NASA

borough, England, the year before he was selected into astronaut Group 5, in April 1966. Once on the astronaut team, he served in back-up and support duties for Apollo 9 and Apollo 12. **Worden** was a little unusual among astronauts in that he did not develop an interest in flying until becoming adult.[24.4] Furthermore, he was something of a poet!

Figure 4.41. The Apollo 15 Crew: **Irwin** and **Scott** seated on an engineering model of the Lunar Rover that they would drive around the landing site on the Moon, and **Worden**, who would remain in lunar orbit in the Command Module *Endeavour*.
Credit: NASA

The main objective of Apollo 15 is summed up by looking at Figures 4.44 and 4.45. They would be landing in the Lunar Apennine Mountains, and visiting the sinuous Hadley Rille. To do this, they needed to use the Lunar Rover (Figure 4.45). **Irwin** described the liftoff on July 26, 1971:

"I heard the word 'ignition' and I sensed, felt and heard all the tremendous power that was being released underneath that rocket beginning to lift me off the Earth. It was a moment of just supreme elation, complete release of tensions. There were tears coming down my face that morning."[14.4]

They headed off for the Moon. Later, he recorded:

"The Earth reminded us of a Christmas tree ornament hanging in the blackness of space. Seeing this has to change a man."[14.4]

Eventually, they reached Moon orbit, and Irwin noted:

"The colors varied with the sun's angle. At the terminator the surface was dark grey, like molten metal. As we traveled toward the sun, the surface lightened to brown, then light tan, and almost white directly below the sun."[14.2]

The two craft separated, with **Scott** and **Irwin** descending to the surface in *Falcon*. **Scott** recounts:

Figure 4.42. The Apollo 15 LMP **Jim Irwin** would occupy the right-hand console of the Lunar Lander *Falcon* during the July 30, 1971 touch-down in the Lunar Apennine Mountains.
Credit: NASA

"Flying the lunar module is a very demanding task. It's the toughest flying job—and I've flown a lot of stuff—the toughest flying job I've ever had."[20.1]

Worden remained in *Endeavour* and carried out a busy routine of conducting science experiments from lunar orbit. **Worden** reported that the science from

Figure 4.43. Al Worden (*right*) jokes with his Apollo 15 Mission Commander **Dave Scott**. **Al Worden** would be the first astronaut to experience an EVA between the Earth and Moon, exiting *Endeavour* during the return to Earth to retrieve equipment from the instrument bay of the spacecraft's Service Module.
Credit: NASA

orbit helped choose the site for Apollo 17 to land.[24.4] He was very comfortable being alone in lunar orbit:

"That is isolation. You could only tell where the lunar horizon was [when at the dark side] by the starlight that it cut off. Otherwise, you couldn't see anything. Total isolation. I thought it was great!"[24.1]

He even deployed a sub-satellite into lunar orbit while he was up there, and sent messages to Earth:

"Hello Earth—Greetings from Endeavour"

in Arabic, Chinese, French, German, Greek, Hebrew, Italian, Russian, and Spanish![24.3] On landing, **Scott** called out:

"Oh, the beauty! The spectacular beauty,"[20.1]

then, as he stepped onto the surface:

Figure 4.44. The magnificent feature Hadley Rille winds its way across the lunar terrain. This was a prime target for Apollo 15.
Credit: NASA

"There's a fundamental truth to our nature: Man must explore."

Irwin later reported these thoughts:

"When I got on the Moon, I felt at home. We had mountains on three sides and had the deep canyon to the west; a beautiful spot to camp[14.4] *... it was the ultimate desert."*[14.1]

Scott and **Irwin** then went to sleep before taking their first EVA. Not at all sure that I could have done that.

Irwin noted, on descending from the LEM, that they had landed on a slope—his first words were:

Figure 4.45. Hadley Rille, in the Hadley–Apennine region, from ground level in July, 1971. **Dave Scott** with the Lunar Rover at the edge of the Rille, which is a mile wide at this point, although it may not seem so. It is very hard to estimate distances on the Moon, because there is no atmosphere to alter the perception of distant points, and there are of course no familiar-sized landmarks, such as trees or houses.
Credit: NASA

"Boy, that front pad is really loose, isn't it?"

Luckily, it was firm enough. **Irwin** and **Scott** deployed experiments, collected rocks, assembled the rover and gave it its first drive on the Moon. **Irwin** recounts:

"The rover really seemed to be another spacecraft, even though we were operating on the surface of the Moon. Every time we'd hit a rock or a bump, we'd just fly into space. So, I estimate we were floating through space a good bit of the time."[14.1]

Eventually they reached the Rille. **Irwin** reported:

"I can see the bottom of the Rille. It's very smooth,"

and **Scott** added:

"the far side has got all sorts of debris."[1.11]

They found it very hard to gauge distances. They encountered a rock balanced on top of a boulder with crystalline properties, named it the Genesis rock, and brought it home with them. **Irwin** commented that wearing the lunar pressurized spacesuit was not easy:

"Every moment in the suit was work. Just opening the hand to grip something was a struggle."[14.1]

He noted:

"We worked too hard on the Moon. We lost essential electrolytes, such as potassium."[14.2]

Scott put things in perspective:

"If you look at the whole expanse of the Plain, there is nothing there ... but that one teeny little thing that's called a lunar module, all by itself. Boy, you're a long way from home. It sort of gives you a perspective of your own dimension in things, which is very small."[20.1]

When they returned to their Lander *Falcon*, **Scott** conducted a science experiment to provide educational interest. He simultaneously dropped a hammer and a (falcon) feather, to show that, in the absence of an atmosphere, they would both reach the surface at the same time, which they of course did! They took off to rendezvous with **Worden** in lunar orbit:

"The ascent was spectacular,"

records **Scott**.[20.1] During the return to Earth, **Worden** was the first astronaut to conduct an EVA in that isolated location between the Moon and Earth (Figure 4.46).

  Worden had done lots of training regarding how he was actually going to do this EVA.[24.4] He recounts:

"You could see the Moon and the Earth at the same time."[24.4]

It was for him an amazing highlight of his journey, and left quite an impression:

"Until you get out there, and you see the star field, and you understand that there is so much out there, I don't care what probabilities you give me, if there's any probability at all of life out there, there's gonna be lots of it. Because, I mean, this thing goes on forever ... [24.1] *Now I know why I'm here ... Not for a closer look at the Moon, but to look back at our home, the Earth."*[24.2]

The rest of the return journey was uneventful, except for a failure of one of the three landing parachutes. Just one of those little things that might have killed them, but didn't. On their return, the crew became embroiled in something that at the time was called "the Stamp Scandal", but which at this degree of hindsight seems an irrelevance. It simply involved the crew carrying too many souvenir postal envelopes to the lunar surface. The crew was recovered by the *USS Okinawa*. From a scientific point of view the mission of Apollo 15 had been a huge success. **Irwin**, however, had developed an irregular heartbeat

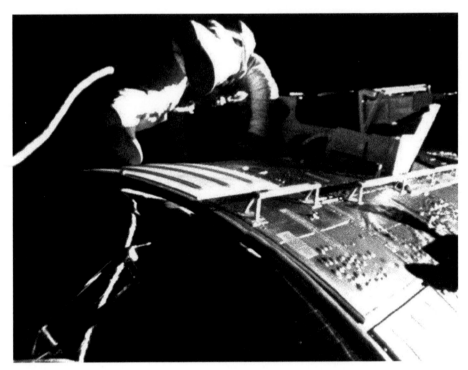

Figure 4.46. **Al Worden** conducting an EVA during the return home from the Moon on the Apollo 15 mission. He retrieved a film package from the Equipment Bay of the Apollo spacecraft *Endeavour*'s service module.
Credit: NASA

during the journey. He did subsequently die of a possibly related heart attack in 1991.

A new postage stamp (Figure 4.47), recorded the use of the Lunar Rover, as the successive Moon landings began to explore more territory.

By this time, it was becoming clear that there would only be two more Apollo Moon landing missions, Apollo 16 and Apollo 17. The highly experienced Commander that Deke Slayton had lined up for the Apollo 16 mission was **John Young** (Figure 4.48). This would be his second flight to the Moon, although of course he had not landed during his Apollo 10 journey. The remainder of the crew would consist of two new guys, **Mattingly** and **Duke**. They are new to us only in that this would be the first spaceflight for each of them, although we have heard about each of them before.

Thomas Kenneth Mattingly (Figure 4.49) was introduced to you earlier in this chapter as a member of the prime crew of Apollo 13, before he was removed due to a measles scare. The big irony in this Apollo 16 crew assignment, is that it was his fellow crew member **Charlie Duke** who had been exposed to the measles resulting in **Mattingly**'s re-assignment at the time of Apollo 13. He had a great sense of humor and reported later:

Figure 4.47. It had been 10 years since **Alan Shepard**'s first US space flight. Now the program had not only landed on the Moon, but was even having astronauts driving around the lunar surface on their own rover vehicle. The US Post Office celebrated this feat.
Credit: Author's collection

Figure 4.48. A light-hearted Apollo 16 crew at press conference—**Young**, **Mattingly**, and **Duke**.
Credit: NASA

"I finally got rid of my hostility for doctors!"[16.4]

Mattingly had been in a kind of friendly competition, which he clearly lost big-time, with his friends from Group 5 (**Mitchell** and **Haise**), about who would get into space first.[17.10] He was the spacesuit specialist amongst the Group 5 astronauts,[13.2] and it was he who named the Command Module *Casper*, after a

Figure 4.49. Ken Mattingly finally has his Moon flight, after being removed from the Apollo 13 prime crew just a few days before launch. For Apollo 16 he was the CMP.
Credit: NASA

friendly ghost in a children's story. Again, they were trying to maintain the interest of the public, and children in particular. **Mattingly** would now get his chance to go to the Moon, and even fly solo in lunar orbit (Figure 4.50).

We met **Charles Moss Duke** (Figure 4.51) when he acted as the CAPCOM at

Figure 4.50. **Mattingly** rides solo in lunar orbit, listening to Berlioz's *Symphonie Fantastique*. The Apollo 16 Command Module *Casper* viewed from the Lunar Lander *Orion*, containing **Young** and **Duke**.
Credit: NASA

the moment of man's first landing on the Moon, during Apollo 11. He afterwards often quipped that he was more famous for his few words at that time than for subsequently walking on the Moon himself.[10.4] **Charlie** came from Charlotte, North Carolina, and was a Lieutenant Colonel in the USAF. He was actually a Naval Academy graduate who went into the Air Force.[10.4] He flew F-86s in Ramstein, Germany, during the Cold War, and had become an instructor at Edwards when he was introduced into the astronaut corps as part of Group 5 in April, 1966. **Duke** (Figure 4.52), served in a back-up capacity for Apollo 10 and Apollo 13, and had also provided some support on Gemini 11 and 12.

Figure 4.51. Charlie Duke, as Apollo 11 CAPCOM, just after the momentous first landing on the Moon. **Duke** responded to **Neil Armstrong**'s: *"Tranquility Base here. The Eagle has landed"* with: *"Roger, Tranquility. We copy you on the ground."*
Credit: NASA

The mission of Apollo 16 was to take place in the Cayley Plains, in the Descartes region of the Moon's southern highlands (Figure 4.53). **Charlie Duke** provides some ironical perspective on the Saturn 5 liftoff:

"At liftoff my heart was pounding 144 beats per minute. John, on his 4th flight into space, was 70."[10.7]

Looking back at the Earth during TLI, **Duke** observed:

"The Earth was a jewel of beauty hanging in the blackness of space."[10.7]

He was much taken by the deep blackness of space:

"It was a texture. I felt like I could reach out and touch it. It was so intense. The blackness was so intense."[10.8]

On reaching lunar orbit, **Young** and **Duke** entered *Orion*, and then separated and descended to the surface, with **Mattingly** keeping lonely vigil in lunar orbit in the Command Module *Casper*.

Figure 4.52. Now, with Apollo 16, **Charlie Duke** gets his own chance at a Moon landing. Lunar Module *Orion*, carrying **Young** and **Duke**, touched down on April 20, 1972.
Credit: NASA

Figure 4.53. The mountainous terrain of the Apollo 16 landing site in the Descartes region is very different than had been the case with Apollo 11. The Lunar Rover is parked high up on a slope (at right) as the crew collect samples; its gold thermal covering providing the only color to the landscape. The lander *Orion* is many miles away down the slope. Credit: NASA

Mattingly loved it! He said:

"There is nothing as personally exhilarating as being solo in a spacecraft on the back side of the Moon. It's the most exhilarating thing in the world. To be there, by yourself, totally responsible for this thing, Dead quiet. And this spectacular, unreal world. Nothing could be more exhilarating. Seeing the Moon in Earthshine is like flying over snow-covered terrain."[16.1]

He describes further:

"I was lying there looking out the window as we moved across the terminator. I was listening to Symphonie Fantastique, and it was dark in the spacecraft. I was looking down at dark ground, and there was Earthshine."[16.3]

But, of course he had to admit:

"Every Command Module pilot envies the people that go to the surface."[16.4]

Meanwhile, the *Orion* Lander had reached the surface:

"Right on schedule, a little blue light came on, which meant that we had made lunar contact. We shut down the engine and dropped down the last few feet. We were elated, of course,"

said **Duke**.[10.10] At 36, he became the youngest of those who walked on the Moon.

Young was excited about the science tasks ahead. Addressing the Moon on his descent down the ladder from the LEM, he said:

"Apollo 16 is gonna change your image!"

Duke's own first words on the surface were:

"Fantastic! Oh that first foot on the Lunar surface is super!"

Young and **Duke** then deployed experiments, collected rocks, and got the Rover ready. It was possible to pick it up and turn it around.[10.7] **Duke** reports that:

"We had a color TV camera mounted on the front of the car controlled by an engineer in Mission Control. He controlled the camera while we just went on about our business."[10.7]

Young pointed out the difficulties of wearing the pressure suit:

"There are a lot of difficult things about wearing a pressure suit. Your hands get awfully tired. By the end of the EVA's you can barely move your fingers,"[25.1] *"and the dust gets into everything."*

Then he added:

"That's the other thing about the Moon, you have no feeling of depth perception. Because you have no telephone poles or anything to relate to."[25.1]

Duke noted the same thing:

"Big objects, faraway, look very similar to smaller objects close in. We jogged, and we jogged, and we jogged ...,"[10.7]

as they tried to reach what they eventually named "House rock." He noticed that the soil had:

"very good bearing strength—you never sank in more than a couple of centimeters."[10.7]

They drove the rover up a feature called Stone Mountain, and **Duke** exclaimed:

"You just can't believe this view!",

followed by **Young**:

"It's absolutely unreal!", [1.11]

and we can see and confirm that sentiment for ourselves with the spectacular image in Figure 4.53. On the Lunar Rover, **Duke** reported:

"The ride was very rough and wild. The back kept fish-tailing; it was like we were on ice all the time." [10.10]

He went on:

"The view of the Lunar Module, Rover, and our flag was always special. Other than the grey and white of the Moon, this was the only color. What a contrast to the stark Moon." [10.2]

Young later commented:

"Even the craters have craters up there!" [25.5]

Duke brought a family photo with him to leave on the Moon (Figures 4.54 and 4.55). It was a symbolic though important gesture, and it survived just long enough for him to take the photo, because, as he records:

"At this point, the Moon's surface temperature was about 230 degrees Fahrenheit. Within a minute or two it was just a shriveled little ball." [10.7]

Duke also left a special medal, celebrating the 25th anniversary of the USAF, on the surface, while saying:

"Happy Birthday, Airforce." [10.12]

After completing their EVA tasks, **Young** and **Duke** re-entered the Lander *Orion* for liftoff from the Moon. **Duke** recounts:

"The floor of the Lunar Module was covered with Moon dust. When we got back into orbit, all the dust floated up into the spacecraft. Mattingly said 'You're not coming in here!' We had to use a vacuum cleaner before he would allow us back into his Command Module!" [10.7]

Mattingly was indeed perturbed when his colleagues came back to lunar rendez-vous after they had spent their time on the Moon:

"You know, I got this all squared away, and it's clean now, and goddang it, I want it to stay that way. Did you want somebody to come back and invade your space? And be dirty? Yeah, you get protective about that!" [16.1]

The final special event of the mission was the EVA during the return to Earth. **Mattingly** conducted this task (and nearly lost his wedding ring in the process, as it floated away toward the open hatch). He became the second of only three men to do this trans-lunar EVA. This is how he described the experience later:

"You had to turn your body to see the Earth, and then the Moon. There's a thing

Figure 4.54. **Charlie Duke** leaves a photo of his family on the Moon, as he completes his surface activities for an April 23, 1972, liftoff and return to Earth. On the back of the photo, Duke had written: *"This is the family of astronaut Charlie Duke from planet Earth, who landed on the Moon on April 20, 1972."*
Credit: NASA

that's the size of an orange, and that's one of them, and there's one over here, and it's a crescent, and it's not quite so big, but that's all there is!"[16.1]

Young subsequently reported about the dangerous EVA work:

"Mattingly left his fingerprints in the handrail."[25.5]

Figure 4.55. The **Duke** family portrait is left on the Moon as a personal memento. In the photo are **Duke**, his wife Dorothy, and children Charles and Thomas.
Credit: Courtesy of Charlie Duke

They returned safely, and were recovered by the *USS Ticonderoga*.

The Apollo program is now almost over. There had been some jockying for the last crew slots. Deke Slayton originally had arranged for an all test-pilot crew, but there was pressure to include a scientist for this last mission, so the chosen crew would carry a geologist, who would replace Joe Engle, who had originally been going to be in the crew (Figure 4.56). **Gene Cernan** would be the Apollo 17 Mission Commander, making his second trip to the Moon. The remaining crew members, both rookies, will complete the roster of 24 guys who went to the Moon.

Harrison Hagan ("Jack") Schmitt (Figure 4.57) was the geologist selected from the Group 4 scientists for the final Apollo mission. He had waited 7 years for the opportunity, after the group was announced in June, 1965, and had been working hard throughout the whole time. **Schmitt** came from Santa Rita, New Mexico and geology ran in the family, with his father being a mining geologist. After obtaining his PhD in Geology from Harvard, he had been a geologist with the US Geological Survey when selected into the astronaut corps. The first thing **Schmitt** needed to do was to learn how to fly. So he was put on a 1-year course at flight school, learning to fly Cessna 172s, T-38s, and then helicopters, including instrument flying. After that, **Schmitt** performed as

Figure 4.56. The Apollo 17 prime crew with Rover in front of Saturn 5; **Schmitt**, **Cernan**, and **Evans**.
Credit: NASA

CAPCOM for Apollo 11, and back-up crew duties on Apollo 15. Because of his special skills, he also performed lunar orbit flight planning for Apollo 8 and contributed to the surface activities planning for Apollo 11.

Ronald Ellwin Evans (Figure 4.58) came from St. Francis, Kansas, and was a US Navy Captain. He flew 7 months of combat missions over Vietnam from the *USS Ticonderoga*. He was a combat flight instructor on the F-8 at the time of his selection to the astronaut corps in April, 1966. **Evans** had maybe the most faithful roster of support duties in the whole program, having served as back-up crew on Apollo 1, Apollo 7, Apollo 11, and Apollo 14, and as CAPCOM for Apollo 7, Apollo 9, Apollo 11, and Apollo 14.

Teamwork was very important in Apollo crews. They spent endless hours in simulators together for years. As **Schmitt** observed:

"The most important part of training was the development of mutual confidence."[19.11]

He also favored working and playing together, spending lots of time with ground crew personnel in the local eatery the "Singing Wheel."[19.11] It would be hard to provide a better example of teamwork in an Apollo crew than that

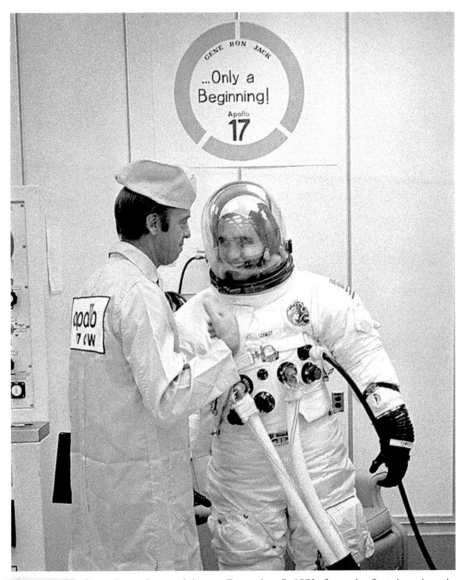

Figure 4.57. Some last-minute advice on December 7, 1972, from the first American in space, **Al Shepard**, to the man who would be the last to step onto the Moon during the Apollo era, **Jack Schmitt**. The sign in the background is a wry commentary on the last Apollo mission to the Moon.
Credit: NASA

seen in Figures 4.59 and Fig 4.60, when **Ron Evans**' widow Jan took part in a Smithsonian celebration giving as good as she got from her late husband's flight crew buddies: "It was a piece of cake! I never saw any one of you willing to hand over to your back-up crews!", she roundly declared.

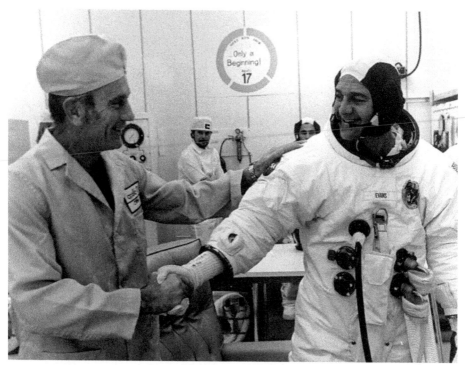

Figure 4.58. Deke Slayton, the astronauts' boss, gives a farewell handshake to **Ron Evans** after he has suited up for the last Moon mission, Apollo 17. This would be **Evans'** first and only flight into space. Slayton was one of the original seven Mercury astronauts, but had to wait until July 1975 before he got his own first and only flight, after the Moon missions were over.
Credit: NASA

Figure 4.59. Janet and **Ron Evans**—in the astronaut limelight in the 1960s.
Credit: NASA

Figure 4.60. Jan Evans, widow of **Ron Evans**, still plays her part as a member of the Apollo 17 team at a 30th anniversary celebration lecture at the Smithsonian's National Air and Space Museum in Washington, DC, on March 18, 2003, 13 years after her husband's death. **Cernan** and **Schmitt** complete the squad.
Credit: Author's collection

The mission of Apollo 17, the final trip to the Moon, had the Taurus Littrow Valley as its destination (Figure 4.61). **Cernan** and **Schmitt** would spend 75 hours on the Moon, including 22 hours of EVA.[7.1,19.6] Such was the progress from the Apollo 11 landing and its mere two hours of EVA. The launch itself was a night launch, which was spectacular, illuminating much of the Florida peninsula. There had been a launch hold at T-30 seconds for a launch computer problem, and **Schmitt** says he *"went to sleep."*[19.11] Of course, for many contractors at the Cape, this was the end of the road. But there was no lessening of vigilance in conducting the mission. **Schmitt** himself declared that:

"Despite being the last launch, there was no relaxation in the working-level dedication to success."[19.11]

Evans described the launch itself:

"I really wasn't sure the crazy thing was going to stay together. Even to read the gauges was almost a guess."[11.1]

They reached Earth orbit, and then headed off Moonwards—the last time this would be done for half a century. **Cernan** later commented on the visual impact of leaving the Earth on the TLI trajectory:

"You say, Hey, I'm out here 150 or 200 thousand miles away from home, going in the other direction. I mean, you have really left society."[7.5]

Schmitt reported later:

Figure 4.61. The Apollo 17 Lunar Lander *Challenger* is the tiny distant object in the center of the image at the landing site on the Moon's Taurus Littrow Valley.
Credit: NASA

"The biggest problem photographers have in printing pictures from space is actually finding a way to print black, absolute black."[19.11]

Finally, they made it to lunar orbit. **Evans** described what he saw:

"There are changes in the color of the Moon as you traverse from sunrise to sunset. At sunset you think the Moon is brownish. Then lighter brown. Then over to high noon and you look out and its bright, bright. Really, really bright."[11.1]

He added:

"As I orbited the Moon, and the Moon was in turn making its twenty-eight-day

orbit around the Earth, I could watch the Earth change from three-quarters, to one-half, and on down to a crescent."[11.2]

Cernan and **Schmitt** climbed into the LEM, departed from the Command Module *America*, and landed the *Challenger* at the appointed place,

"with dust streaming away from the descent engine,"[7.1]

leaving **Evans** to his thoughts as he continued to orbit the Moon. **Cernan** went down first to the surface, followed by **Schmitt**. So, **Schmitt** was technically the last man to set foot on the Moon. They conducted three separate EVAs, deployed experiments, and collected rocks as they traveled across the lunar surface on their Rover. **Cernan**'s first words on the surface were:

"Oh my golly! Unbelievable; but it is bright in the sun."

Schmitt's first words on the surface were more prosaic, even comical. Let history take note that what he said was:

"Why don't you come over here, and let me deploy your antenna?"

Yes, it's true. You can Google it! **Schmitt** reported what we have heard already from earlier moonwalkers:

"You were never going to have enough time to do what you wanted to do,"[19.1] and: *"All experiment deployments constituted a challenge in the suit. All deployments took far more time than planned."*[19.2]

On the rover, he recalls:

"You kept bouncing off the surface. The rover left the surface at every bump."[19.1]

A fender on the Rover broke off, and they effected a repair using a piece of the Flight Plan and duct tape. **Schmitt** comments:

"Had we not fixed the fender, keeping cooling surfaces on the Rover and on our helmets free of dust would have been a major and time-consuming problem."[19.2]

He added, regarding the difficulties in finding your way:

"The clarity brought on by a lack of atmosphere gives the impression that objects are closer than they really are. This made it difficult to estimate distance, so I used the known distance of my shadow and any given sun angle to calibrate my estimates of near field distances and crater diameters."[19.2]

The decision to include a geologist paid dividends when **Schmitt**,[1.11] discovered some orange soil (Figures 4.62 and 4.63):

"There is orange soil. It's all over! Orange! I've got to dig a trench, Houston!"

Schmitt got on board the *Challenger*, followed by **Cernan**. So **Cernan** was the

Figure 4.62. The color code on the geological tripod, or gnomon, clearly indicates that geologist **Schmitt** has found orange soil amongst the lunar regolith of Taurus Littrow.
Credit: NASA

last man to leave the Moon—that is, as he named his book *The Last Man on the Moon*. **Cernan** later commented:

"We're not going to be back here for a while. What should we be doing? Is there something we should say, or do, or be?"[7.5]

In fact, the last words which he said on the Moon were:

"Let's get this mother out of here … Engine start: push."[7.1]

On the way home **Cernan** developed his thoughts on the wonderment of it all:

Figure 4.63. Back at the LRL, in Houston, the orange soil can be clearly seen.
Credit: Collectspace.com

"I'm looking at something called space that has no end, and at time that has no meaning."

He also noted:

"we spent most of the way home discussing what color the Moon was."[7.6]

Schmitt began providing weather reports to Earth using 10-power binoculars.[19.11] **Evans** conducted the last of the three cis-lunar EVAs when he left the spacecraft to retrieve film canisters:

"I was having a ball ... you're not really a spaceman when you are in the confines of your spaceship ... you go outside[11.1] *... I was making a visual inspection of what was out there. I'm an engineer, I'm doing my thing. So, what do I get to look at? The urine dump!"*[11.1]

Schmitt later reported:

"We raised more questions than we answered. You always do. That's the nature of field geology."[19.1]

He said that the lunar surface was:

"a pitted and dusty window into our past[19.10] *... Apollo 11 began the process of understanding the evolution of the Moon and, indeed, the early evolution of the*

Figure 4.64. A geologist, and the last man to step on the Moon. **Jack Schmitt** back in the Lunar Module *Challenger* after one of his lunar EVAs, with much evidence of the dusty lunar surface on his space suit.
Credit: NASA

Earth. From the Moon we gained information about the early history of the Earth that would have been impossible to gain from Earth itself."[19.14]

Figures 4.64, 4.65, and 4.66 capture the crew coming back from the Moon. **Ron Evans** would have the particular pleasure of being recovered by his old carrier, the *USS Ticonderoga*, which had already served that function for Apollo 16.

We have now followed all the Moon missions, and all of them are summarized in Appendix C.

Figure 4.65. Gene Cernan, covered in Moon dust, was the "last man on the Moon," because he followed his colleague **Jack Schmitt** up into the Lunar Module *Challenger* after the last EVA.
Credit: NASA

What did they all do later? We shall talk about that in the next chapters, but none of this crew flew in space again. So, we have completed the chapter on the Moon missions, and have now encountered all 24 men who went to the Moon. No doubt, you have noticed how different they were at the time. Their subsequent lives following the Moon missions would provide further information on how this experience affected them. Meanwhile, Figure 4.67 offers a neat framing of the Moon landings, with the Commanders of Apollo 11 and Apollo 17, the first and last missions, at their joint *alma mater*, Purdue.

The next chapter will address the remaining spaceflights of any of the 24 Moon travelers who had still some space flying to do. It was indeed a hard act

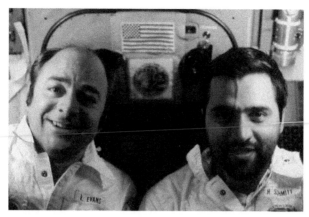

Figure 4.66. The journey home in Command Module *America*—**Ron Evans** and **Jack Schmitt** on Apollo 17. The return to Earth came on December 19, 1972, the last time a crew has returned from the Moon.
Credit: NASA

Figure 4.67. First and Last. **Neil Armstrong** and **Gene Cernan** were the Commanders of Apollo 11 and Apollo 17, the first and last of the 6 Apollo missions which resulted in landings. The photo was taken at a ceremony at their *alma mater* Purdue University in West Lafayette, Indiana on October 27, 2007, at the dedication of the **Neil Armstrong** Hall of Engineering.
Credit: news.uns.purdue.edu

to follow, and many of the Apollo astronauts retired at the end of Apollo 17. However, some stayed on. There were Skylab missions. There was the joint Soviet-US mission ASTP, and there was the beginning of the Space Shuttle program. The last spaceflight of the last of the Apollo astronauts to fly into space, would take place 12 years after the last Moon landing. I call this next chapter "The Tail End."

CHAPTER 5

The Tail End

THEIR LAST FLING IN SPACE

Only 6 of the 24 Moon travelers remained actively involved in spaceflight beyond the end of the Apollo missions, 8 if you count those who continued to act on back-up teams. The period concerned covers the 12 years from January 1973 to January 1985. Importantly, the massive era of funding at 5% of GDP which was in evidence through the 1960s in support of the Moon program was over. It had tapered off even before Apollo 17 flew. But **Conrad**, **Bean**, **Stafford**, **Haise**, **Young**, and **Mattingly** were not done. Even though the budgets were severely reduced, there were nevertheless three new programs to interest them: Skylab, the Apollo–Soyuz Test Project (ASTP), and the Space Shuttle. The first two used left-over Apollo program hardware. During most of the period discussed in this chapter, during the tail-end of the Apollo program and the onset of the Space Shuttle flights, **John Young** became the Chief of the Astronaut Office at Johnson Space Center, Houston, Texas, carrying on the tradition which had been started by **Alan Shepard** and Deke Slayton. **Young** would be astronaut chief from 1974 to 1986.

What could possibly motivate astronauts, who had already been to the Moon, to continue to put their lives at risk in the cause of the space program? What could possibly motivate the public, who were funding all NASA ventures, to want to continue doing so? The first attempt at addressing these questions was Skylab.

Skylab was a space station, proposed by von Braun, which was created from a re-use of the Saturn 5's third stage, since it was not going to be needed to get back to the Moon. The stage was fitted out with experiment racks and a living area, and contained a vast empty space at its core which could be used for various free-space, zero-g experiments. It became a valuable way to explore long-duration flights and their impact on the human body. Skylab was visited by three crews in succession, each arriving and departing using left-over Apollo capsules. Eventually, the 48-ft long, 85-ton craft was de-orbited and burned up on re-entry, although some large chunks did reach the Australian outback.

The first crew to rendezvous with Skylab was led by veteran **Pete Conrad** (Figure 5.1). His fellow crew members were rookies Paul Weitz and Joe Kerwin. Kerwin had been one of the Group 4 scientist astronauts who joined

Figure 5.1. Crew of first Skylab mission—Weitz, **Conrad**, and Kerwin, May 25–June 22, 1973.
Credit: NASA

in June 1965. He had waited 8 years for this opportunity to conduct science in orbit. Paul Weitz was from Group 5 (April 1966), amongst which group nine of his peers had been able to fly to the Moon. The crew of this mission knew that the station had been damaged during launch (Figure 5.2), and were able to make some temporary repairs, by means of a parasol sunshade which they deployed, thus saving the habitat from being destroyed due to runaway thermal control issues. They had, in effect, saved the orbital outpost, so that it could continue to be used for laboratory experiments. **Conrad** always subsequently regarded this as his most important mission, demonstrating what only a human crew's ingenuity could achieve in the realm of space. Even without the missing second solar wing, they were able to remain in space for almost a month, conducting scientific experiments.

The second Skylab crew was also led by an Apollo Moon traveler, in this case it was **Alan Bean**, in his first spaceflight as Mission Commander. His other

Figure 5.2. The Skylab space station, made from a remaining Saturn 5 third stage no longer needed for a Moon mission. The first Skylab crew installed the golden "sunshade" when there was damage to the station during liftoff, that had removed its original protective thermal cover, and caused the loss of one solar wing. There would be three crewed-missions to the station.
Credit: NASA

Figure 5.3. Crew of the second Skylab mission, July 28–September 25, 1973—Garriott, Lousma, and **Bean**.
Credit: NASA

crew members were rookies Owen Garriott and Jack Lousma. Garriott was one of the scientists from Group 4, and Lousma was one of the "Original Nineteen" of Group 5 (Figure 5.3). The crew spent almost two months in space aboard the facility, conducting science experiments in the laboratory. Some 35 years later, Garriott would be able to watch his son Richard head off into space to visit another orbiting space station—the International Space Station—with Richard flying as a space tourist, in 2008.

There was a third visit to Skylab, with a crew not containing any Apollo Moon travelers. The all-rookie crew for that mission of 84 days, starting in November 1973 in the orbiting laboratory, was Carr, Gibson, and Pogue. Gerald Carr and Bill Pogue were from Group 5, and Ed Gibson was the last of the Group 4 science astronauts to fly ($8\frac{1}{2}$ years after joining the astronaut team). The crew conducted many experiments, particularly related to solar physics. Despite these missions, Skylab never captured the attention of the general public—at least not until the station was finally de-orbited on July 11, 1979 and there was general concern over where the pieces might land.

The last fling of the Apollo spacecraft was the ASTP. This was a remarkable exercise in statecraft—one could call it the dawn of space diplomacy. The idea was to have craft from each of the two Cold War states, the USA and the Soviet Union, rendezvous in space. This would have benefits in terms of potential future space rescue procedures. Of course it would also mean that US astronauts would be able to visit the latest Soviet spacecraft, Soyuz, in orbit;

Figure 5.4. Crews of the joint US–Soviet ASTP mission, July 15–July 24, 1975. From left to right, Slayton, **Stafford**, Brand (due to fly in the Apollo spacecraft), Leonov, and Kubasov (to pilot the Soyuz spacecraft), all behind a model of the combined spacecraft as they would be connected in orbit. Deke Slayton was aged 51 when he achieved this, his only spaceflight, and he had to learn Russian in order to do it.
Credit: NASA

and likewise the Soviets could fly in Apollo. There was a period of preparations when both the crews would need to visit the training facilities of their opposite counterparts. Language would be an issue. The crews (Figure 5.4) undertook to learn each other's language, and so it became necessary for the US crew of Mission Commander **Stafford**, Slayton, and Brand to learn Russian. Similarly the Russians—Leonov (who had conducted the world's first spacewalk) and Kubasov—had to learn English (or Oklahoman, as Leonov subsequently pointed out!).[22.7] As back-ups for the mission, **Alan Bean** was the Commander, supported by **Scott** and **Evans**. Vance Brand had been one of the Group 5 astronauts recruited in April 1966, and would be a rookie for the flight, acting as Command Module pilot. He had, during those previous 9 years, served on many back-up teams, including Apollos 8, 13, and 15. Deke Slayton, since there was no Lunar Module (LEM) on the mission, became the one and only Docking Module pilot. Slayton's heart murmur issue had gone away after

Figure 5.5. Space Shuttle *Enterprise* is released from its Boeing 747 carrier aircraft over Edwards Airforce Base, as part of the landing test program, in October 1977.
Credit: NASA

treatment consisting of very high doses of vitamins. When **Stafford** entered the space program, Slayton was his boss. I asked **Stafford** about the role reversal and he replied:

"Deke wanted to have command, but he had not flown, and I had 3 missions plus a lot of rendezvous experience, and so they put me in command. We never had any problems with that."[22.7]

In Figure 5.4, the docking module, which was specially designed and built for the mission, can be seen as the black part on the model between the Apollo spacecraft (*left*) and the Soyuz (*right*). It effectively consisted of a tunnel with docking mechanisms on each end, and pressurization systems to balance the pressure and atmospheres of the two spacecraft. The two crews became very good friends, and in particular **Stafford** and Leonov continued their friendship for the rest of their lives. **Stafford** said:

"I have come to know Alexei Leonov like a brother. I did not have any brothers or sisters."[22.7]

Leonov would later prove instrumental in getting the first space tourists trained for their flights on board Soyuz spacecraft. For two days the craft remained docked in orbit, and there were many ceremonial activities and broadcasts performed. There was an accident during the recovery of the Apollo spacecraft, when nitrogen tetroxide thrusters continued to fire, and gas leaked into the cabin. **Stafford** quickly reacted and made sure the crew wore oxygen masks, as they waited inside the capsule for recovery. The remaining members of the two crews get together for commemorations,[22.9] and were inducted into the Astronaut Hall of Fame at the Air and Space Museum in San Diego in March 2001, where I witnessed the continuing conviviality of these former Cold War foes.

Thus truly ended the Apollo era. The focus was now on the development of the Space Shuttle. This was intended to bring benefits of re-usability to human spaceflight. Many more astronauts would be recruited to fly on Shuttle missions, some of which would carry 8 crew members. But for our story, we are only concerned with the last flights of the three remaining Moon travelers **Haise**, **Young**, and **Mattingly**. These Apollo veterans each contributed their considerable expertise in getting the Space Shuttle program into operation.

We begin with **Haise**. As a matter of fact, he did not fly in space again, but he did fly five of the critical Approach and Landing Tests (ALTs) of the Shuttle program to ensure that, on returning from space, the Shuttle orbiter would be able to safely maneuver and land on a runway. In Figure 5.5 we see the Shuttle orbiter *Enterprise* separating from its carrier aircraft to begin one such ALT.

Fred Haise flew five of these ALT missions, as Commander, with Gordon Fullerton as his Pilot in the right-hand seat, between June 18 and October 26, 1977 (Figure 5.6).

So **Haise** laid the groundwork for the missions to follow. The Space Shuttle would become the workhorse for the next stage of human spaceflight, carrying US and international crews for three decades until being retired in July 2011. During that period there would be two catastrophes causing loss of crew (*Challenger* in 1986 and *Columbia* in 2003). It never achieved its cost targets and was ultimately deemed too dangerous to fly. Certainly its early test pilots took enormous risks.

The crew observed in Figure 5.7 were to conduct perhaps the boldest test flight of any in the space program, when they flew the first Shuttle space mission STS-1. **Young** and Crippen took off vertically from Kennedy Space Center on April 12, 1981, and landed two days later horizontally on the runway at Edwards in California. I recall watching the TV in amazement when they landed after their spaceflight, ostensibly just like a regular aircraft— because crews of all previous spaceflights had returned to Earth by parachute into the ocean. Amazingly, there had been no previous unmanned powered flight of the Shuttle into space before **Young** and Crippen undertook the mission. Nor had there even been a test of the abort procedures. **Young**

Figure 5.6. The test pilot crew of the Space Shuttle ALTs—**Fred Haise** and Gordon Fullerton—walking away from a successful landing in the Shuttle *Enterprise*, October 26, 1977, at Edwards in the Mojave Desert. The 747 carrier craft is seen flying away (*top right*). This had not been a space flight, however, since the flight, and the *Enterprise*, were only used to explore the regime after re-entry and the subsequent atmospheric handling of the Shuttle orbiter.
Credit: NASA

determined that such a test would be more dangerous than heading straight to orbit:

"Let's not practice Russian roulette, because you may have a loaded gun there."[25.5]

There had been an accident during a Countdown Demonstration Test when three pad workers died due to inhaling pure nitrogen due to a procedures oversight. Furthermore, during the flight, the crew—Crippen was a rookie, but very experienced in Shuttle software[25.5]—reported the scale of the noise and vibra-

Figure 5.7. Crew of the first all-up test of the Space Shuttle, STS-1, Commander **John Young** (*left*) and Pilot Bob Crippen, seen during training in the mission simulator. The space shuttle *Columbia* would make its first flight into space taking off on April 12 and landing on April 14, 1981.
Credit: NASA

tion problem, and the flight sustained resulting damage to thermal protection tiles (16 were lost—fortunately in non-critical areas). They also learned about the Shuttle's aerodynamic behavior on re-entry and how it differed significantly from the computer models. **Young** made reference, from orbit, to the pad deaths that had occurred while making the flight ready. In fact, in retrospect, the crew were lucky to escape with their own lives. However, with this risky test flight, they had verified that the Shuttle was "space-worthy," and a new era could thus begin (Figure 5.8). By pure chance, and as a result of a series of delays leading up to the launch, STS-1 took off precisely on the 20th anniversary of Yuri Gagarin's first manned spaceflight. Hence, April 12 has become doubly significant among dates to celebrate spaceflight achievement.

There were another two Shuttle test flights, in November 1981 and March 1982, with crews that did not contain any Moon travelers, and for each of them **T.K. Mattingly** was back-up Mission Commander. In these test flights, corrections were introduced to address the problems that **Young** and Crippen had identified.

Mattingly got into space again as Commander of the fourth Space Shuttle mission, with Hartsfield as his pilot (Figure 5.9). The flight was the final one before the Space Shuttle was declared fully operational, and was therefore the

Figure 5.8. Space Shuttle *Atlantis* launches a crew into orbit during the three decades of new generation spaceflight.
Credit: NASA

last to only carry a crew of two. Furthermore, from this flight onwards, NASA would no longer continue with Deke's process of assigning complete back-up crews to each mission, a signal of moving on to a different, and indeed operational, phase of spaceflight, using a fleet of five re-usable aircraft-like spacecraft (*Atlantis, Challenger, Columbia, Discovery*, and *Endeavour*). For this mission, **Mattingly** and Hartsfield did carry a few scientific and military payloads, before returning to Earth to a July 4th patriotic welcome (Figure 5.10).

The landing at Edwards was attended by President Reagan and his wife Nancy, who were clearly delighted to have the Shuttle return to their home state. Reagan used the occasion to announce his new space policy, which effectively recognized that the Shuttle program was now moving from development to operations. Soon afterwards, assembling the International Space Station (ISS) would become a major focus of the space program, and only the Space Shuttle would be able to transport the larger components to do that task.

Young and **Mattingly** still each had one spaceflight to go. From November 28 to December 8, 1983, **John Young** would fly the last of his six space missions, as Commander of Shuttle *Columbia* for STS-9. **Young** later reported,

Figure 5.9. Ken Mattingly gets into space again as Shuttle Commander of *Columbia* for mission STS-4 from June 27 to July 4, 1982. His pilot is Henry (Hank) Hartsfield, who had been one of the CAPCOMs during his flight in Apollo 16, a decade earlier.
Credit: NASA

somewhat whimsically, that:

"I got to fly the first set of re-used solid rocket motors."[25.5]

Over 18 years, he had flown twice in Gemini, twice in Apollo, and twice on the Shuttle. On this, his last spaceflight, **Young** flew a mission which carried in the Shuttle's payload bay the European Space Agency (ESA) Spacelab laboratory, to conduct microgravity research. Also on this flight he flew the Shuttle's first of many international crews (the crew included Ulf Merbold, ESA's first astronaut). The 6-member crew also included NASA astronaut scientist Garriott, whom we discussed earlier with Skylab. The Shuttle continued to be dangerous, however, and **Young** had to deal with problems—which could have been fatal—with on-board flight computers crashing and a fire in a rear engine compartment during landing. In later years, I heard him give a talk at Washington DC's National Air and Space Museum, while he was still running the astronaut team at NASA, where he said:

"My wife would kill me if I would ever fly into space again!"[25.5]

He had survived a dangerous and lengthy career at the leading edge of manned space flight, in an era when, in his own words:

"We didn't do reliability calculations."[25.5]

Figure 5.10. Shuttle crew of **Mattingly** and Hartsfield deliver their craft *Columbia* to their Commander in Chief, and Mrs Reagan, landing at Edwards Air Force Base, California, fittingly on the July 4, 1982.
Credit: NASA

Mattingly would prove to be the last of the tail-enders. He flew one more time, in January 1985, as Commander of Shuttle mission STS-51-C in *Discovery*, delivering a Department of Defense military satellite payload into orbit. His flight would also, in retrospect, underline the dangers of the Shuttle design. **Mattingly** and his pilot Loren Shriver took off in very cold temperatures, and subsequent analysis of the solid rocket boosters, which had been returned by parachute for refurbishment, revealed that the seals between the rocket motor segments had been badly damaged. Almost exactly a year later, on January 28th, 1986, there would be the *Challenger* disaster, when the whole crew (including teacher Christa McAuliffe) would perish in the destruction during launch of their Shuttle, due to a low-temperature launch making the O-ring seals between booster segments brittle. So, **Mattingly** also had ended his spaceflight career with a lucky escape. Sometimes, even the skills of an exemplary test pilot cannot be enough to overcome the systemic problems of any particular spacecraft design.

All of these "tail-end" missions of the Moon travelers have been summarized in Appendix C.

The Shuttle continued flying until July 2011, and the program even survived one other disaster, of *Columbia* on February 1, 2003. This time the destruction was due to damage to the thermal protection tiles through vibration of the whole vehicle stack during liftoff, causing particles of foam to be dislodged from the External Tank hitting the Shuttle orbiter's protective surfaces at high speed. The damage was not realized until the fateful re-entry resulted in the craft's break-up due to thermal stresses. **Young** had experienced this phenomenon in a survivable form on the first flight of the Space Shuttle, and even though a great deal of effort was applied to solve the problem, it was never satisfactorily resolved. During its years of operation, there would be many triumphs for the Space Shuttle fleet, throughout its 135 missions carrying 833 crew members—such as the launch and subsequent repair missions of the Hubble Space Telescope, and the building, block by block, of the International Space Station (ISS). The crew selection criteria for astronauts were gradually changed. For example, women started being recruited into the astronaut corps in January 1978. The first US female astronaut, Sally Ride, flew on the Space Shuttle on June 18, 1983, and the first female Space Shuttle Commander, Eileen Collins, took her crew into orbit in July 1999. The Shuttle was used for commercial space missions between 1982 and 1986, when communications satellites were deployed from its payload bay, and the author negotiated for the launch of one such satellite deployment, which however was eventually canceled following the *Challenger* disaster. The three surviving space shuttle orbiters, *Discovery*, *Endeavour*, and *Atlantis*, were consigned to space museums in 2012. As this book is being written in 2017, we await the next generation of spacecraft for manned spaceflight, some of which are intended to operate as re-usable space taxis to carry astronauts—as well as space tourists—into orbit.

We have now come to the end of Section I—The Way It Was. We have followed the spaceflight careers of each of the 24 astronauts who took part in the flights to the Moon. It is now time for some reflections on that Golden Age of space travel. And so we start Section II—Post-Apollo Reflections.

SECTION II

Post-Apollo Reflections

6 AFTERWARDS

7 POST-FLIGHT ANALYSIS

8 CONCLUSIONS

"There can be no thought of finishing, for aiming at the stars, both literally and figuratively, is the work of generations, but no matter how much progress one makes there is always the thrill of just beginning."

Dr. Robert Goddard, letter to H.G. Wells, April 1932

(quoted in *Aiming for the Stars* by Tom Crouch, 1999)

CHAPTER 6

Afterwards

It was all nearly fifty years ago. What has happened in the intervening years to the men and materials of Apollo? We start with the materials.

The launch vehicles were generally expendable, and so there is not much left from the vehicles that were used in the Moon missions. There are, of course, remnants of all expendable first stages in the Atlantic Ocean. Jeff Bezos, the boss of Amazon and the Blue Origin space transportation company, even recovered one of the rocket engines from the Apollo 11 launch, and it is on display temporarily at the Museum of Flight in Seattle before moving to a new gallery, "Destination Moon," at the National Air and Space Museum (NASM) to open in 2020. We can see complete un-flown Saturn 5 vehicles in three places: Alabama, Houston, and the Kennedy Space Center (KSC). Also, the descent stages of the six Lunar Landers remain to this day on their respective landing sites on the lunar surface. We can see them in high-definition imagery from lunar orbiters operating today. We can even still see the disturbance of the lunar regolith, or soil, from astronaut footprints and Lunar Rover tire tracks. The artifacts on these historic sites are technically the preserve of the Smithsonian Institution, so someday in the distant future we can look forward to seeing a "Smithsonian Museum on the Moon." In the nearer term, we may get to see them again from high-definition images sent back to Earth from the craft being sent by teams competing in the Google Lunar XPRIZE, and for which "heritage preservation" on the Moon is an important aspect of their plans (see Appendix D).

The part of the spacecraft that carried the crews of course returned to Earth, and for each of the missions the Command Modules are now in museums in various places around the world. Figure 6.1 indicates where Apollo 11's *Columbia* ended up. This was the only bit of the enormous Saturn 5 rocket stack which returned to safe harbor. There is other Apollo hardware elsewhere in the US (in California, Texas, Alabama, Kansas, and other states), and also in the UK, Japan, Switzerland, New Zealand, Canada, and France. Even the Smithsonian cannot of course show you the Lunar Lander from the first Moon landing, but they have a flight model on display, which would have been used if missions had continued beyond Apollo 17. As a volunteer docent at the NASM, the author enjoyed presenting these artifacts to the public, and noted that at this point the general public had forgotten a great deal of the history of these

Figure 6.1. The spacecraft ended up in museums, such as the National Air and Space Museum (NASM) of the Smithsonian Institution in Washington, DC. In the image we see (*from left to right*) the Mercury Capsule *Friendship Seven*, the Gemini 4 capsule, and Apollo 11's Command Module *Columbia*.
Credit: Author's collection

missions. The artifacts were a way of bringing back the consciousness of what it was like to be bolted into any of these craft and rocketed off to places unknown. The museums also retain many of the spacesuits which were used, but these present a major problem to the conservation professionals whose duty it is to preserve them for future generations. The problem in a nutshell is that they were not designed to last for decades, or half-centuries. There are many layers of rubber that are slowly disintegrating, and so the suits must be kept in a special environment to prevent further decay.

The story is not so good when it comes to the original facilities which were used during the Moon launches of the 1960s. Figures 6.2–6.4 evidence the demise of the original Mission Control room at Cape Canaveral. The building that housed this facility was in a poor state of repair, and the contents of the room were re-constituted in a Visitors Center at KSC, when it was demolished. In many cases, the humidity and salty sea air of the Florida facility guaranteed that the huge structures that were used during the Apollo era would not survive beyond the first half-century after they had been used (Figure 6.5). It is interesting to note in passing that the subject of the destruction of the launch facilities

Figure 6.2. Mercury Mission Control in its heyday, around 1962. Flight Controller Gene Kranz is seated in the center of the picture, wearing the dark blue waistcoat vest.
Credit: NASA

Figure 6.3. The beginning of the end of the Mercury Control Room at KSC, April 28, 2010.
Credit: NASA/Retrospaceimages.com

Figure 6.4. Final destruction of the Mercury Control Building in May, 2010.
Credit: NASA

Figure 6.5. Demolition and collapse of a launch tower, October 22, 2010. This is at the site that was used for the original Atlas launches, Launch Complex 36.
Credit: NASA

Figure 6.6. A new era begins to establish itself at the Cape facilities. This former Apollo launch pad is now the site for the Falcon missions of the commercial spaceflight company SpaceX at KSC.
Credit: SpaceX

and control rooms was of some concern to at least Deke Slayton and Scott Carpenter from among the early astronauts, and they attempted some intervention which is in the documentary record, but which was however unsuccessful. The good news, however, is that there is an active program of re-purposing some of the early facilities for a new generation of spaceflight. Observe, for example, in Figure 6.6, that Elon Musk's SpaceX organization now uses the launch pad, once used for Moon missions, to launch its *Falcon* vehicles. Other pads are now being used by teams aiming to compete for the Google Lunar XPRIZE (Appendix D).

Space Exploration Technologies (SpaceX) has been developing re-usable boosters that deliver re-usable spacecraft (called *Dragon*) to the International Space Station (ISS). They are on target to start carrying human cargo once the launcher has been "human-rated" for this purpose. Once that has been done, this old Apollo launch site will become the take-off point for new generations of astronauts and space tourists heading into orbit, and even possibly eventually to the lunar vicinity.

So much for the materials. What can we report about the men who remain? Well, as we write, many of them are still alive. They continue to deal with "life after Apollo," although of course most are totally retired from active participation in anything related to space and the Moon missions now. With 45 years having passed since the last mission, the youngest of them is now 82. They all had to find out what to do after they had achieved the pinnacle of flying to the Moon. Especially as they continued to have honors heaped upon them in the

Figure 6.7. Memorialized in parks or buildings. This is in Cocoa Beach, Florida.
Credit: Visitspacecoast.com

succeeding years. It must have been very hard to return to "normal" life. Indeed, **Al Shepard** distinctly stated that it was impossible:

"I don't think any of us ever returned to normal life; I don't think any of us were normal people to start with."[21.1]

As a sweeping generalization, which we shall soon correct on a case-by-case basis, the remaining 20 or 30 years of their professional lives, after they stopped being astronauts, consisted of boards, consulting, beer distribution, talk circuits, and ministries. I am particularly interested in their views about how they see their contribution, and the future of space exploration. Where do they think we are headed now? Some of the more famous of the 24 have ended up with some relatively permanent markers of their existence. There are places and buildings and roads scattered about that have been named for the early space pioneers. **Al Shepard**, for instance, has this ocean-side park named after him (Figure 6.7) in Cocoa Beach, Florida. It is near the hotel that he and his fellow Mercury astronauts used to frequent, when they were staying at the Cape for a launch.

Other facilities frequently named after astronauts are schools and university buildings. **Aldrin** and **Armstrong** both have this honor. You can find them, for instance, at Shaumberg, Illinois (**Aldrin** Elementary); Gates, New York (**Neil**

Figure 6.8. Some are remembered through ship launchings. **Neil Armstrong**, the first in a new class of oceanographic research ships, was launched on March 29, 2014, and christened by Carol Armstrong.
Credit: CBSNews.com

Armstrong Elementary); and at Purdue University in Cincinnati (The **Neil Armstrong** Hall of Engineering). **Neil Armstrong** also has a NASA facility named in his honor (the **Armstrong** Flight Research Center, at Edwards, California). Surely one of the most magnificent lasting legacies is to have a ship built and launched with your name. This happened with many of the early astronauts from a naval background (including **Shepard**, Glenn, Schirra, and **Armstrong**—Figure 6.8).

Then there is the artwork; there are paintings (including murals), sculptures, and statues, among public works of art. There are murals in such places as the Astronaut Hall of Fame and NASM, some of them by **Alan Bean**, himself a Moon traveler. **Bean** has indeed produced an entire body of artwork depicting his fellow Moon travelers in the years since he left NASA. Regarding statues, we can list at least the following five of them: there is the **Armstrong**, by Fagan at Purdue; the **Bean** at Wheeler, Texas, by Mickey Wells; the **Cernan** at the Kansas Cosmosphere in Hutchinson by Weeks; the **Lovell** by Rottblatt-Amrany at the Adler Planetarium in Chicago; and the **Swigert**, which is by the Lundeens, and is at both the Denver Airport and the US Capitol (Figure 6.9). Of course, in the realm of art, there are also the astronaut autobiographies; untypically amongst astronauts, **Collins** even wrote his own.

Figure 6.9. Some have statues in their likeness. This statue of **Jack Swigert** at Denver Airport, by artists George and Mike Lundeen, is a replica of an identical one in the Capitol Visitors Center, Washington, DC.

Credit: Denver International Airport

After they returned from the Moon, the Moon voyagers all shared various experiences in common. Then of course they had their individual life experiences which we shall soon discover. What they shared in common would ensure that each of them had become an icon, whether they liked it or not. Some of

Figure 6.10. Buzz Aldrin at a book signing for his Mars book at the Ronald Reagan Presidential Library, June 8, 2013.

Credit: acorncapture.com

them were sent on world tours, meeting Royalty and world leaders. Most of them took part in book signings, where lines of the public would wait to have their book(s) autographed and have a word with the famous author. Many of them either received phone calls from the US President, or were invited to meet at the White House, or give a talk to Congress. None of them originally "signed up" for this. They were service officers and pilots of fast military aircraft. There was no preparation for them or their spouses and families in how to handle any of this. We shall now see how each of the 24 did handle the challenge, and how they conducted the rest of their lives. We approach this for each Moon traveler in turn, in alphabetical order.

So we start with **Buzz Aldrin** (Figure 6.10). In some ways he is a perfect example of how all the fame and anti-climax could make life difficult; but he also embodies a resolve and resilience which has made it possible to recover from these difficulties, and continue to this day as a living symbol of the possibilities of human spaceflight.

On returning from the Moon landing, the Apollo 11 team and their wives went on a world tour which was exhausting, and included a visit to London on October 14, 1969, where this author was part of an enormous crowd greeting the Moon travelers on the steps of the American Embassy in London. Other

Moon missions were still continuing when **Buzz** left NASA in July 1971, and the USAF (after a brief period as Commander of Edwards Air Force Base) in March 1972. He had trouble initially on taking stock:

"You gotta come back and do the laundry, and all the rest of that stuff, and face reality. Sometimes, as if it never happened. Or despite it ever happening."[2.12]

He has spent a lifetime dealing with the psychological aspects of his Moon journey,[2.10] and struggling with describing his experiences to a public that always wants more.

"I'm not sure really how the layperson reader is ever going to grasp whatever words are going to try and describe this. I've felt totally inadequate in ever trying to do it with spoken words."[2.12]

It is not as though he has never tried. He is the author of a succession of books about his Apollo era and later years,[2.7–2.11] and he is a frequent face on the conference and TV talk circuit.[2.1,2.2] He is a persistent advocate for continued space exploration, and takes any opportunity, however seemingly remote from space, to continue to spread the word. We have seen **Buzz** on the TV program *Dancing with the Stars*, for instance. But he shines when the subject turns from his historical past, and moves to the future potential of space. He has created a succession of companies and foundations (such as Starcraft Enterprises and Sharespace) and coalitions,[2.4] with the aim of developing future space transportation systems.

"I have great ideas about how to bring about lunar exploration and space transportation, how to bring about Mars exploration ... you do what you are good at, and I realized I'm good at that."[2.5,2.12]

Buzz developed his concept for "Mars Cyclers" in 1984,[2.13] and has continued to refine details since that time.[2.5] He took on a leadership role as chairman of the Board of Governors with Wernher von Braun's National Space Society, and has been a regular attendee at the Society's annual conferences (the International Space Development Conference—ISDC) since then. He is a regular visitor to the White House advocating with each new president in turn for a continued American presence in interplanetary space. **Aldrin** has also been a Commissioner on a blue ribbon Commission exploring America's future aerospace options, and managed thereby to get language about "public space travel" for the first time into such a commission report.[2.4] He is a strong supporter of space tourism and initiatives for public involvement, especially connecting with children,[2.17] in space exploration—such as the Google Lunar XPRIZE (see Appendix D). But Mars, and indeed Mars settlement, is his main focus:

"It was 66 years from the Wright Brothers' first flight to the first landing on the Moon. If we move 66 years from landing on the Moon, that'll put us into 2035. That's my reasonable estimate for permanence on Mars."[2.14]

"Exploration does not mean languishing in LEO ... we explore or we decline."[2.4]

Buzz became Chancellor of the International Space University (ISU) in 2015. I should report in fairness that I am not impartial in my assessment—**Buzz** has been very kind to me, supporting my own endeavors for the Gateway Earth concept,[2.6] even writing a foreword for my space tourism book—*The Wright Stuff*.[2.19] You can, however, come to your own conclusions and watch him in person on two videos containing cameo appearances.[2.16,2.18]

  Bill Anders (Figure 6.11) comes next as we review the post-Apollo stories of our Moon travelers. After returning to Earth following the success of Apollo 8, **Anders** was initially on the back-up crew for Apollo 11, but then

"selected out of astronaut work."[3.2]

For a while he did some NASA space policy work in Washington, being Executive Secretary of the National Aeronautics and Space Council from 1969 thru to 1973.[3.2] He became disillusioned by the politics of the Space Shuttle, however, and reverted instead to his old profession, and became Chairman of the Nuclear Regulatory Commission from 1975 thru to 1977.[3.2] He then served

Figure 6.11. **Bill Anders** continues to enjoy flying in his vintage military aircraft.
Credit: Heritage Flight Museum

a term as US Ambassador to Norway in 1977, before finally homing in on the area which would occupy the bulk of his post-Apollo attention: management of the contractor General Dynamics in various key positions between 1977 and 1994. He completed the Harvard Business School's Advanced Management Program in 1979.[3.2] **Anders** eventually, in 1990, became Chairman and CEO of the General Dynamics Corporation.[3.2] He retired in 1994 to concentrate on his hobbies of sailing his sailboat *Apogee*, running the Heritage Flight Museum with his son, flying his Mustang in air races, and cross-country skiing. He says he has three times turned down the post of NASA Administrator,[3.2] because he believes that NASA is off-track, and needs to undergo a major wind-down of the work of its various centers, since the public is worried about cost. **Anders** had been involved in massive cost-cutting regimes in General Dynamics (following the end of the Cold War), and did not want to go that way again.[3.1] He became comfortably well off during his years with General Dynamics, and this has made him something of a target. Recently (in 2017) he won a court case against his accountant for misappropriation of $750,000 of funds.

Anders is a realist when it comes to continuing the space program. In an interview for the *Seattle Times* in December 2012 he said:

"In my view, contracting much of NASA's former role to private industry is the way it's going to have to be done. That's what made the airlines and the aerospace industry. What happened to the space program? It's finally run into reality. The reality is that people like it, they'll go to a launch, they'll want to get an astronaut's autograph. But they are not willing to pay for it."

I was able to listen to the good-natured ribbing that still takes place between the Apollo 8 crew members,[3.3] when I attended a talk they gave, (Figure 4.9). Often, the ribbing is about who among them took the famous "Earthrise" photograph. But **Anders** has an interesting take on that image, which provides us all with an important perspective:

"My favorite picture was not "Earthrise". It's one where the Earth is sort of blurry and small. I don't think we've ever really gotten it across to people through the photography the perspective of it. That you've got to see all the black, all the nothing ... in order to totally get the smallness, aloneness, insignificance of this pretty ball you're looking at."[3.4]

In the December 2012 *Seattle Times* interview, he said:

"It's ironic . . . we came all this way to discover the Moon, and what we really did discover is Earth."

As was the case with many of the Apollo astronauts, they subsequently saw the former Soviet Union in a new light later in their lives. **Anders** recalls:

"I spent my early Air Force career chasing Russian bombers over Iceland when they invaded Icelandic air space. Later, I was flying a supersonic interceptor that had three nuclear-tipped rockets. In looking back, to think of a junior Air Force

captain flying with these things under the wing kind of boggles my mind, but that was the Cold War. America had a real paranoia."[3.5]

Next up is **Neil Armstrong** (Figure 6.12). This most-famous of the Moon travelers had a wry sense of humor, which he too rarely displayed to the public. A classic example is when a friend said: "I passed by your place last evening," and **Neil** simply replied: *"thank you."* He is generally recalled as a quiet, introverted, and modest engineer. He wrote no autobiography, apart from the official crew account.[4.5] He had a way of rationalizing fear. Listen to how he describes the possibility that the engine of the Lunar Module's Ascent Stage would not fire when attempting to leave the Moon to return to Earth:

"That the ascent engine would not fire was a concern I had for a long period ahead of the flight. Nevertheless, if the engine does not light, you are not in a dangerous position. You have a lot of time. You can talk it over with the ground, decide what kind of alternative methods you're going to try. And you're not out of options at that point. So, compared with a lot of serious things in a flight, that wasn't up at the top."[4.6]

Figure 6.12. Neil Armstrong (*right*) makes a point at a meeting of the NASA Advisory Council on February 8, 2007, with Chairman **Harrison Schmitt** looking on.
Credit: NASA

Talk about "Mr. Cool"! He had also surmised:

"My instinct—not a carefully reasoned statistical study, but my instincts—told me we had a 90% chance of a safe return, and a 50% chance of a safe [lunar] landing."[4.6]

There is a wonderful testament to his calmness in his *Senior Year High School Yearbook*, where under his photo it says "He thinks, he acts, 'tis done!" After his post-Apollo 11 world tour, he initially settled in to a NASA headquarters job, as deputy Associate Administrator for Aeronautics, but was not happy operating "within the Beltway," and he retired from NASA in 1971, setting up a small dairy farm back in his home State. However, he would find that you can never really leave all that behind, once you have been a Moon traveler.[4.1] He had met Lindbergh several times,[4.2] and learned from him to be circumspect in dealing with the demands of a public which can never be satisfied. Having first seen him during his post-Apollo 11 world tour, I encountered **Armstrong** again in February 2007,[4.4] when he was acting as a member of the NASA Advisory Council, under the Chairmanship of **Jack Schmitt**. In that committee he was concentrating on his own preferred area of enduring professional interest, namely aeronautics, and I was able to provide him with some space tourism data to help in his analysis of Air Traffic Control (ATC) requirements at the time. He was concerned about the slow pace of developments in ATC, and declared that one reason was that there is:

"a lot of mystery regarding responsibilities."[4.4]

Armstrong had spent about 10 years in relative obscurity as Professor of Aerospace Engineering at the University of Cincinnati (1971 to 1979), leaving because he found academic politics difficult.[4.3] For a while, he was rather uncomfortable working as a Chrysler spokesperson.[4.8] Later, in 1986, he was recalled to the aid of NASA when appointed as Vice Chair of the Presidential Commission investigating the break-up of the Shuttle *Challenger*. Then, he spent a decade (1982–1992) as the relatively invisible Chairman of an aviation consulting firm, Computing Technologies for Aviation, of Charlottesville, Virginia. He died, aged 82, on August 25, 2012, from complications of a heart bypass operation. His enduring insights into the long-term future of planet Earth underscore his concerns for future ongoing space exploration.

"Earth looks like it could not put up a very good defense against a celestial onslaught."[4.3]

"The Earth is not benign. I have seen shooting stars below me. I have seen the violence of nighttime thunderstorms like giant mushrooms illuminated by ferocious lightning. I have seen gigantic hurricanes with enormous winds."[4.8]

Armstrong went further:

"We can conjure up a hundred reasons for migration from Earth: change in our atmosphere, overpopulation, radiation growth, nuclear holocaust, disease, collision

with a comet or asteroid. Earth's magnetic polarity changes from time to time. There is evidence suggesting that another reversal is possible sometime soon. We don't have the foggiest idea of what the effects of such an event might be.[4.8]

His conclusion from all this potential misfortune?

"We must eventually rise above our differences and become a true family of nations."[4.8]

He died without being able to figure out what was needed, to make space settlement elsewhere a possibility. He said, rather ruefully:

"I'd like to see us on Mars, but it's too expensive."[4.3]

As if to underline the great diversity among the ranks of the Moon travelers, it would be hard to find a much bigger difference than that between **Armstrong** and our next candidate **Alan Bean** (Figure 6.13). **Bean** himself would point that out later, when he recounted the tale about **Armstrong**'s ejection from the Lunar Landing Training Vehicle in May 1968. They shared an office and secretary in the astronauts' building at Johnson Spaceflight Center in Houston. This is how **Bean** tells the story.[4.7] He came back from lunch and found **Armstrong**

Figure 6.13. **Alan Bean** enthusiastically interacts with visitors to his art exhibition at the National Air and Space Museum on July 16, 2009.
Credit: Author's collection

at his desk as usual, just shuffling some papers, and was surprised to learn from corridor talk that **Neil** had just ejected from the Lunar Landing Training Vehicle (LLTV) at the last moment an hour previously.

"So I go back to the office. Neil looked up, and I said 'I heard that you bailed out of the LLTV an hour ago.' He just said 'I lost control and had to bail out of the darn thing'."

Bean was astonished that **Armstrong** did not make a big deal out of it, as most other astronauts would have done; **Armstrong** simply later recorded:

"That is true, I did go back to the office. I mean, what are you going to do? It's one of those sad days when you lose a machine."[4.2]

We have seen how **Bean** hung on to astronaut work until he had been Commander of a Skylab mission, and then back-up Commander for the joint US/Soviet mission ASTP in 1975. After that, he retired from the US Navy in October 1975, continuing with NASA as interim Chief Astronaut, doing some preparatory work for the Shuttle program, before finally retiring from NASA in June 1981, to devote himself full time to creating artworks.

Bean had begun to be interested in art back in his Pax River days but, as a military test pilot "jet jock," did not discuss his interests much with his colleagues. It was having successfully completed his journey to the Moon that transformed him, and gave him a new direction in life.

"Since that time I have not complained about the weather one single time; I am glad there is weather. I've not complained about traffic; I'm glad there are people around. Why do people complain about the Earth? We are living in the Garden of Eden ... I've been really happy since the Moon trip. I've felt that I got a lot of luck in life to be able to do that and go to the Moon ... Mostly it made me have a lot of courage to do what I wanted to do, and be happy about it. That's one thing that really allowed me to be an artist. I wouldn't probably have had the courage to be an artist."[5.8]

On a scale of courage rating from 1 to 10, how many of us would place "being an artist" up there with "getting on a rocket and blasting off to the Moon"? **Alan Bean** always had a limiting view of his role as a naval officer, stating:

"I knew how to be a lieutenant, but not a colonel."[5.4]

He often seemed to be seeking approval rather than leading, and he was not politically active either within his field or in general, claiming:

"I was too busy to consider external socio-political events."[5.4]

Bean has described, however,[5.6] how he learned much about attitude, teamwork, and leadership from **Pete Conrad**. And as a consequence he has become a motivational speaker, asserting with confidence that:

"Nobody in this room has any idea about what they could do with concentrated effort for five years."

Moreover, he did have a very focused visual acuity, and was very single-minded about his art. He came to realize that he had the skills and interest to do something that nobody else could do. He was going to

"focus on art—the things that I could do something about [5.4] *and leave a body of work"*

which recorded the experience of being on the Moon from the point of view of someone who had been there.[5.3] **Bean** has also left a good record of his views in person as a participant in a series of DVDs of the Moon travelers.[5.12–5.14] As a professional artist he has been very successful and some of his original paintings of the Moon missions fetch prices upward of $100,000. He has also published books of his artworks to access a wider audience, and limited edition prints for those of us who cannot afford the price of his originals. **Bean** wanted to offer a visceral connection of the Moon experience to those who saw his artwork, and so he includes imprints from his Moon boots, his lunar geology hammer, and his lunar core-tube sampler, as part of each painting's background texture, and even the odd grain of Moon dust from the surface coating of his badges.[5.5,5.6] The magic of Moon dust even extends toward making his used paint brushes desirable objects among collectors of memorabilia. His artist's eye has provided some wonderful insights into the experience that we might not have otherwise had. Listen to him, in the movie *For All Mankind*, describe the difficulty of conveying the experience of being 240,000 miles away to those of us on Earth who only can see the photos of the Earth that the Moon travelers brought back:

"Whenever you see a photo of the Earth [that we brought back], there's always a frame, something that holds the Earth up. But in space, there's nothing holding it up."

Alan Bean does not have, or at least express, any strong views about the future of humans in space, or settlement. He does care, however, that children understand the importance of the historic lunar exploration events.[5.3] The most I ever heard him declare on the topic of the direction, or means, of future space travel was in Washington, DC in 2003, when after listening to a pitch I had given about private space flight at a conference, he did say to me:

"If space tourism can be part of helping young people realize their dreams about space, then that's good."[5.2]

He prefers to let his work speak for itself.

Perhaps one of the most determined of the Moon travelers who wanted to leave the experience behind and "move on" has been **Frank Borman** (Figure 6.14).

"I left NASA in '70 and never looked back,"

Figure 6.14. **Frank Borman** recounting the story of Apollo 8 at a 40th anniversary talk at the NASM in Washington, DC, November 13, 2008.
Credit: Author's collection

he claimed.[6.1] What does the record show? Was he in fact successful in that resolve? **Borman** always was a highly focused military officer, and perhaps he never changed. He always considered that:

"After Apollo 11, the Mission was over,[6.1] *. . . I had no interest in Moon rocks."*

He considered that

"it was not worth the risks [just] to pick up rocks."

He certainly could not get interested in the Shuttle:

"Where's the mission?[6.1] *Apollo did not die,"*

he made clear.[6.4]

"It accomplished exactly what it set out to do. It was a program of limited scope.

It's one of the strengths of our society that the exceptional becomes ordinary and then forgotten. It's a strength because we're always looking for the next thing."

He has lamented:

"Back then, we had a goal, the money and the support of the country."[6.1]

What did he actually do after returning from his Apollo 8 Moon mission? He continues to have strong religious views to guide him, and is an Episcopal lay reader, recognizing human frailty:

"The only person that hasn't made a mistake was crucified about 2,000 years ago."[6.1]

What must have been quite a challenge to his frame of reference was when he took part in supporting the joint Soviet/US ASTP mission.[6.1] He had previously been motivated to defeat the Soviets in the space race to the Moon, because he

"took the Cold War seriously ... and did not want to be bested by a second rate communist country."[6.1]

I was in a crowd of many hundreds of people squeezed into Downing Street in London, outside the residence of the British Prime Minister, when **Borman** visited the UK in February 1969. This was to be the first of many foreign liaison trips that he would undertake as Special Presidential Ambassador,[6.1] a kind of goodwill ambassador to Europe and Asia. But he had left both NASA and the USAF before 1970 was over. He had found his new focus: the airline business. His love of aircraft had begun at age 15, he owned his own aircraft and he took his leadership skills, including graduation from Harvard Business School's Advanced Management Program in 1970, to Eastern Airlines, where he would remain from 1970 to 1986, variously as Chairman of the Board, President, and CEO. **Borman** would find that flying to the Moon was a doozie compared with managing Eastern Airlines during the 1978 deregulation era, and his many struggles with the unions at the time are well documented in his book *Countdown*.[6.3] It is a very different thing to manage in a military environment, where

"clear leadership made decisions,"[6.1]

to operating with the multiplicity of sometimes conflicting objectives to be faced at the top of an airline. After an exhausting time, he left Eastern in 1986, and thereafter became Chairman and CEO of the Patlex Corporation, based in Las Cruces, New Mexico, where his son owns the car dealership **Borman** Autoplex. Patlex, founded in 1986, concerned itself with patent protection for laser patents. **Borman** remained at the helm of this perhaps oddly suited firm, until in 1996 at 68 years old he retired when Patlex was bought out by Database Technology.

Borman does not continue with his connections to the space program, saying:

"I lose it ... colonizing the Moon and all that other baloney ... I'm too practical. I don't think that's ever going to happen, except maybe as a scientific base."[6.1]

But not even **Frank Borman** can get away forever from his Moon exploits, and that's why we see him make a cameo appearance in a DVD,[6.5] and why, see Figures 4.9 and 6.14, he turned up to help re-tell the story at a 40-year anniversary of Apollo 8. He declared, after reminding the 75-year-old Chairman of General Dynamics, **Bill Anders**, that he was still a rookie, that:

"We are all three lucky to have had the support of our wives, and I am proud to say we are the only Apollo team where every member still has their original wife."[6.2]

While **Frank Borman** claimed that he:

"left NASA and never looked back,"

Gene Cernan (Figure 6.15) lived and died as Mr. Apollo Moon Missions personified. He never tired of telling the tale,[7.1] and in fact wrote one of the best books about the Apollo experience *The Last Man on the Moon*.[7.4] He was able to always view the flights in a wide and even grand context, and was an excellent communicator. He was however much frustrated in later life by the US and NASA's perceived inability to continue with the human exploration missions that had characterized the 1960s, focusing on the way that the entire country had become *"risk averse."*[7.3] Having twice risked his own life on flying to the Moon, we can understand his perspective. But he was unable to understand, and therefore support, the shift in leadership and focus of the space program, once Apollo (and its associated 5% of GDP funding) had ended. He kept lobbying on the Hill for a new *"Kennedy moment,"* when (he hoped and demanded) a new President would lead the charge to Mars, and the funding would follow. He declared:

"The Shuttle is not enough [to get public attention][7.3] *... The public will care about going to Mars."*

Cernan made some of his more memorable assertions in his book, various presentations, and TV commentaries:

"I know the stars are my home. I learned about them, needed them for survival in terms of navigation. I know where I am when I look up at the sky. I know where I am when I look up at the Moon; it's not just some romantic abstract idea, it's something very real to me. See, I've expanded my home."[7.7]

His Apollo experience carried religious or at least spiritual significance for him:

"The Earth was not created by accident"[7.3]

was his oft-repeated conclusion from having observed it from the distance of the Moon. He continued in later life to pursue his love for horses, hunting,

Figure 6.15. **Gene Cernan** in San Diego on March 24, 2001, as part of a Hall of Fame investiture.
Credit: Author's collection

fishing, and flying.[7.3] But after a period of doing some ASTP liaison work in 1973, when he was the senior US negotiator with the USSR (with some CIA links),[7.3] **Cernan** would not be able to make any further significant professional contributions to space exploration, beyond offering general advice such as *"Keep it simple."*[7.3]

Cernan left NASA (after doing some early Shuttle simulator work), and the US Navy, on July 1, 1976. He initially considered going into politics in Illinois or Texas,[7.3] but finally gravitated to the oil industry, which abounds in the Houston area which was his home. He became VP of Coral Petroleum,[7.3] which he referred to as energy management consulting. But this proved to be unsatisfying. He found it a struggle to discover anything that would give him the same kick as beating the Soviets to the Moon. He then, in 1981, tried aerospace consulting,[7.3] during which time he helped start a small airline and was flying Learjets, at 51,000 feet, for the Canadian firm Bombardier. In 1994, aged 60, he became Chairman of the Board of Johnson Engineering, adjoining his old astronaut HQ of Johnson Spaceflight Center. He had open heart surgery in 2005, and subsequently died on January 16, 2017, of heart-related complications, aged 82.

Cernan had done some TV work for ABC,[7.3] and he left some great interviews as a legacy on various DVDs,[7.7,7.8,7.10] and in his own biographical movie *The Last Man on the Moon*.[7.9] I had tried to get him interested around the time of the Wright Brothers' centenary, in space tourism,[7.2] but he really was not going to be persuaded that spaceflight should be carried out by anyone but highly experienced military test pilots. In that respect, he was following the lead of his hero-buddy, and former project Mercury astronaut, Wally Schirra. One of my prized possessions is a copy of **Cernan**'s book *The Last Man on the Moon*, dedicated from him to Schirra, with the words:

"With respect and admiration. Thanks for showing us the way! Geno."

In another copy, he provided me with the personal inscription:

"Keep up the space business!"[7.2]

so perhaps he was not so far from being convinced of the one true way ahead- which is, of course, commercial!

Alan Bean, we have seen, became a professional artist after his Moon travels. **Mike Collins** (Figure 6.16) also enjoyed painting, but saw himself as a recreational artist who occasionally sold a painting. **Collins** is very relaxed and self-deprecating in nature, with a great sense of humor. In 2009 he declared:

"If I had to sum it up at this stage of my life, in one word, I'd just say 'lucky'."[8.3]

Although he was never the activist that his fellow crew member **Buzz Aldrin** became, he nevertheless firmly believes in space exploration.

"Space is the only physical frontier we have left ... and its continued exploration will produce real, if unpredictable, benefits to all of us who remain on this planet."[8.4]

Where should we be going? *"Mars, Mars, Mars and Mars,"* he emphatically suggests.[8.6]

Collins devoted some considerable care, after the "official" Apollo 11 book was published,[8.1] to capturing his thoughts in his own two books *Carrying the Fire* and *Liftoff*,[8.4,8.5] and therefore does not readily answer mindless enquiries:

Figure 6.16. A contented **Mike Collins** shows images of some of his watercolors at an art show in Tucson, Arizona.

Credit: Novaspace/Author's collection

"I have had it with certain questions, and maybe the way they are asked ... I have written two books, and convinced myself, for better or for worse, that I have said everything I want to say. I've just written this 400-page book telling them what it was really like up there, and they say 'Oh, I loved your book—now tell me what was it really like up there?'."[8.2]

Charles Lindbergh clearly appreciated the effort and care that he put into writing his books, and wrote the foreword for *Carrying the Fire*, saying "books such as this by **Michael Collins** stimulate the mind, enhance awareness, and assist us on our way." His writing conveyed much of what it was like, and how risky the venture really was.

"I was surprised that a Saturn 5 never blew up,"[8.6]

just about sums it up.

Now that I have dissuaded you from asking **Collins** to tell you what it was like, all over again, I do therefore recommend you read what he did say in his books, because the language was not written by a ghost-writer—they are his own distilled thoughts. First of all, what does he say about the Moon?

"It was a totally different Moon than I had seen before. The Moon that I knew from old was a yellow flat disk, and this was a huge three-dimensional sphere, almost a ghostly blue-tinged sort of pale white. It didn't seem like a very friendly or welcoming place. It made me wonder whether we should be invading its domain or not."(8.8)

And then his contrasting view of Earth:

"The Moon is so scarred, so desolate, so monotonous, that I cannot recall its tortured surface without thinking of the infinite variety the delightful planet Earth offers: misty waterfalls, pine forests, rose gardens, blues and greens and reds and whites that are missing entirely on the gray-tan Moon."(8.7)

He expresses frustration with the inadequacy of language and photographs to show the lunar travelers' perspective to the Earthbound:

"Seeing Earth on an 8 × 10-inch piece of paper is not only not the same as the real view, but even worse—it denies the reality of the matter. To actually be 100,000 miles out, to look out four windows and find nothing but black infinity, to finally locate the blue-and white golf ball in the fifth window, to know how fortunate we are to be able to return to it—all these things are required."(8.7)

As the author of this present book, I have tried, in my turn, to convey something of the enormity of the undertaking by presenting, in Figure 1.1, an image which was not available until 35 years after the last Moon-landing mission. I hope **Mike** will approve—I believe so. **Mike Collins** enjoys fishing as well as his amateur watercolor painting. He does not stray onto **Al Bean**'s territory, however, and keeps to subjects which reflect his joy at being back on Earth. He is a *plein air* painter, as is my wife, and we were delighted to find that he was capturing the delights of rural Maine on Deer Isle at his easel at the time we were honeymooning there in September 2005. I shared this anecdote with him at a Tucson convention in 2017, and so we now have one of his New England watercolors—"Deep Cove"—in our collection. He has supported the connection of art and space exploration for a long time, and wrote the Foreword for the book *NASA Art: Fifty Years of Exploration.*

Mike left NASA in January 1970, and became the first Director of the Smithsonian Institution's National Air and Space Museum (NASM), on the Mall in Washington, DC., as well as Undersecretary for the Smithsonian Institution. His former Gemini 10 Commander, **John Young,**[25.5] gave him high marks for:

"bringing in the Air and Space Museum [construction] inside the schedule and under budget—something that has never been performed by any government agency!"

When I was a volunteer docent guide for a few years around 2010, many of my colleagues remembered, with affection, the era when **Collins** had run the show. He recalled those years when, with his Apollo 11 crewmates, at the 40th anniversary of the first Moon Landing, he gave a lecture in the NASM's IMAX lecture hall to an overflowing audience. He said:

"We never designed it for an event like this."[8.3]

Collins was able to ensure that his Apollo 11 Command Module *Columbia* had pride of place in the entrance hall to the museum. What was not so well known, however, was that he had inscribed his own personal message inside the module,[8.6] as soon as they were safely aboard the rescue carrier, reading:

"Spacecraft 107—alias Apollo 11, alias 'Columbia.' The Best Ship to Come Down the Line. God bless her. Michael Collins, CMP."

After he had completed his term at the NASM, he finally handed over protection of his precious craft to his successor, and **Collins** took a low-visibility job from 1980 to 1985 as an executive director with an aerospace company, as VP for Vought Corporation, in Arlington, Virginia. He eventually retired to Marco Island, Florida, and

"now only worries about finding a really good Cabernet under ten dollars."[8.3]

He leaves a great personal account of his experiences in cameo appearances in a DVD.[8.9] On the 40th anniversary of Apollo 11, he made this remark, cementing his place in history:

"I think Apollo was a dividing line, putting Earth, after four billion years, into a new category, the Big League of planets. To me, that was the most significant thing about Apollo."[8.3]

If Mike Collins has a low-key and relaxing personality, then **Pete Conrad** (Figure 6.17), was the opposite. He died on July 8, 1999, aged 69, as a result of

Figure 6.17. **Pete Conrad** enjoying being part of a "newspace" conference in Los Angeles, November, 1997.

Credit: Author's collection

crashing his Harley on a mountain road in California. He had also raced cars in Florida with **Al Worden**.[24.4] **Pete** was described as "colorful," and indeed outside the Johnson Space Flight Center in Houston, Texas, there is an avenue of trees which are illuminated with bright lights to commemorate the early astronauts. All the lights are white except for those illuminating **Pete**'s tree.

"I didn't have some great worldly philosophical bullshit about going to the Moon ... It was just, you know, Goddamn, this is gonna be neat!"[9.1]

Alan Bean always regarded **Pete** as his mentor:

"My favorite story about Pete was about him letting me fly the lunar module during the lunar rendezvous phase. I said 'Well, the ground might not approve ...,' and he says 'Don't worry about it. We're behind the Moon. They can't tell what we're doing'."[5.8]

Some wag pointed out that on Apollo 12, they lost a huge chunk of time on the surface simply by laughing.

Conrad had served on the Skylab mission after the Moon missions ended (and even regarded his Skylab 1 mission, when he and his crew had effectively saved the outpost, as more significant than his Apollo 12 Moon trip), but decided to leave NASA rather than wait for the Shuttle program to get started.[9.5] So he left NASA and the US Navy in 1973, and initially went into the Cable TV business. He was COO of American Television and Communications Corporation (ATC) from 1973 to 1976, and he used to be a regular on a San Diego channel that interviewed owners of light aircraft about their rides. He also took part in Amex commercials. He was still, in 1996, going for records, and he flew a Learjet around the world in just over 49 hours—the jet is on view at Denver Airport. The time it took to circumnavigate Earth did not of course compare with having done it repeatedly in Earth orbit in 90 minutes. He began to realize that his space work was not yet over. He still had contributions to make. He had noted that the good old days of the space program were over, and the Shuttle was not going to meet the needs for future space exploration in the Conrad mold, continuing:

"Our future in space is at a critical point. To really move ahead, we require an economically viable method of space travel."[9.4]

He became an early supporter of the Space Frontier Foundation,[9.2] which advocated for a new approach to space using commercial motivation to drive space developments, and which viewed the resources of the Solar System as potential commercial targets. He looked back at the Moon, and saw it differently with the benefit of hindsight:

"We only have to look at the data already accumulated from Apollo to realize that the Moon is rich in raw materials of significant potential; providing an alternative energy source as well as a multitude of manufacturing and industrial applications."[9.4]

Conrad became a Vice President of McDonnel Douglas, as they attempted to develop a new kind of launch vehicle which could use a single stage to orbit and be totally re-usable. It was called the DC-X or Delta Clipper, and, for some of its test flights in 1993, **Conrad** acted as the remote-control test pilot. He became Chairman of his own company, Universal Space Lines, saying:

"We are going to be the first operators of the commercial reusable launch vehicle. I don't necessarily want to build it. If I have my way, I'll also go fly it, so that I can go back into space."[9.5]

Just over a year before he died, we shared a lunch table at a conference hotel near Los Angeles airport, when he told me:

"I wanna fly cargoes and passengers on real commercial launches. Anywhere on the planet in 45 minutes or less."[9.2]

Figure 6.18. Charlie Duke and his wife Dotty at a prayer service in July 2009 in New Braunfels, Texas.
Credit: San Antonio Express News

His widow brought out a posthumous book of his life story in 2006.[9.3] The title? *Rocketman*. Going to the Moon did not change him. It only made him *more* **Pete**.

Moving on, we come to the charming and self-effacing **Charlie Duke** (Figure 6.18). He had seen the Apollo Moon trips, and their retrospective view of the home planet, in a very special way:

"Apollo is still to me a great adventure. It's a great step for Man. I think it will go down in history as a reach out to the inquisitiveness, the sense of exploration, the sense of adventure, the quest for knowledge—that Man has within us, and it will always be that to me."[10.1]

He harbors regrets that it is not possible today to repeat the Apollo adventure:

"Apollo took 8 years and 2 months from Kennedy's statement. Today, we couldn't even write the proposal in 8 years and 2 months."[10.7]

He had been the youngest man to walk on the Moon, at age 36, and continues to believe that:

"The space program is very, very important for our country."[10.4]

"You see Spaceship Earth, and we're all down here together. And we need to get

along, and we need to love one another, and with technology we can help solve our problems."[10.7]

Charlie did work for 3 years on the Space Shuttle following his Moon flight,[13.2] but left the Agency in January 1976, before that vehicle was flown; a source of some future regret.[10.4] He did, however, continue in the USAF reserves, as a Brigadier General. He still turns up for commemorative events,[10.5] and talked with me at one in 2003, after a presentation I had given about space tourism,[10.3] sharing his enthusiasm for the idea:

"Oh yes. Sure. They will love it and tell their friends—Hey you gotta do this! It is just a wonderful experience."

He inscribed a photo of his Moonwalk to me in April 2013 with the simple dictum:

"First explorers, then tourists, then settlers!"

After leaving the program, **Duke** went into the private sector, and left aerospace altogether, deriving his income from a mix of real estate businesses (**Duke** Investments, **Charlie Duke** Enterprises) and a Coors Beer distributorship. But these occupations did not define him. Or change him. What did both of these was his late Christian conversion in 1978:

"I began to love my wife and I began to love everybody on Earth."[10.7]

Once he had time to decompress from the intense period of the Apollo program, **Charlie Duke** had realized that he had been totally focused on his life as a jet jock, and not at all on his family.[10.4] **Duke** became a motivational and spiritual speaker, with his wife Dotty by his side, as a result of his re-commitment to a Christian ministry, becoming President of the **Duke** Ministry for Christ. **Charlie** and Dottie co-wrote their testamentary book *Moonwalker* in 1990.[10.6] There is some great video of **Duke** giving cameo appearances.[10.9,10.11] In June 2017, I discussed the Gateway Earth architecture (a method of using space tourism revenues to help fund space exploration—see Appendix E), with him at a Tucson conference. His comment?

"That's a great idea. Hope it works out!"

It is a source of some mild regret that I did not meet all of the folks who went to the Moon. **Ron Evans** (Figure 6.19) is one of the Apollo Moon voyagers with whom I found no opportunity to have an interaction. He died in 1990 from a heart attack, an inveterate smoker, aged 56. However, as I described earlier (Figure 4.60), his widow continued to stand up for him amongst the light-hearted banter of his former Moon traveler crewmembers 13 years later. As we recorded in the previous chapter, **Evans** remained with NASA right through the Skylab and ASTP missions, and even into early Space Shuttle developments, but did not fly again after his Apollo 17 mission. He retired from NASA eventually in March 1977, taking a position as an energy corpora-

Figure 6.19. Ron Evans during a post-Apollo 17 trip to New Delhi, India, in July, 1973.
Credit: Photodivision.gov.in

tion (coal industry) executive. He had been the proverbial "good soldier," having taken back-up crew positions all the way from Apollo 1 to the US–Soviet ASTP.

"We all had the best job in the world … we really did."[11.1]

Something which affected him very much as a military pilot was the fact that the Vietnam War was taking place at the time of the Apollo flights.

"I was in a position where I could really do something for my country … and I did not have to be fighting a war."[11.1]

Consider how much his life had changed because of his Moon flight:

"We flew Apollo 17 in December of 1972. The POWs [Prisoners of War] were released from Vietnam in January of '73. We happened to be at [legendary singer] Frank Sinatra's place in Palm Springs, watching their return on TV. Some of my friends walked down the ladder after they landed back at Clark Air

Force Base, some after having been a POW for six years. They didn't even know I'd gone to the Moon. I could just as well have been in the [ironically named POW camp] Hanoi Hilton [with them]. And, instead, I'd been selected into the space program, gone to the Moon, and returned [home] the month before they did. I'll tell you, that gets to you."[11.1]

Evans had produced a VHS video cassette about his Moon experiences, titled *Let's Fly to the Moon,* perhaps to mirror Sinatra's iconic song *Fly Me to the Moon.* The return of the vets had been sobering.

In what other ways had the Moon experience changed Evans?

"It is my belief, that no one could have participated in such a venture without reinforcing your belief, in whatever belief you had before you left. In my case, it confirmed there is a God."[11.1]

Evans died before the notion of space tourism had achieved currency, so we cannot know what his views might have been, except for this clue:

"You just cannot relate to it until you've had the experience of being in zero-g. It's absolutely delightful."[11.1]

His time in the coal industry was ultimately not satisfying for him, and he took a position back in aerospace, as a Sperry Flight Systems executive. Other sources of income for him were real estate and public speaking engagements.

Someone else who went to the Moon, and then switched to the energy industry, was **Dick Gordon** (Figure 6.20).

In **Gordon**'s case, he would become an energy consultant (with Energy Developers (1977), Resolution Engineering (1978), AMARCO (1980), and Astro Sciences (1982)), but not until he had first tried out other directions perhaps more in line with perceptions of an old jet jock. **Gordon** was particularly disappointed when Apollo 18 was cancelled, because he was scheduled to be its Commander. He had hoped to make the last stage of the journey from lunar orbit and down to the surface.[12.3] And so he retired from both NASA and the US Navy in January 1972.[12.4]

"It was a great experience that I had. It's one of those things that comes along, not once in a lifetime, but once in a millennium, maybe."[12.1]

His only involvement in space these days is in consulting on space museums, space movies, astronaut conventions (Figure 6.20), and theme parks. He does not "connect the dots" towards the solution to further space exploration, but shows the disappointment that many of the Moon travelers experienced that, nearing the ends of their lives, they do not see signs of significant further space exploration endeavors. He mourns:

"NASA gained a 'how to' attitude from Apollo, but not sure it exists today[12.3] *... Who's out there, challenging humankind to go further and return to the Moon, or go to Mars? The Moon would be a great training ground for Mars."*[12.4]

Figure 6.20. **Dick Gordon** in August 2005 at an astronaut convention in New Jersey.
Credit: Author's collection

Dick Gordon does recognize the possible contribution of space tourism, and he assured me,[12.2] that:

"Any space tourists in lunar orbit would love the views."

But what were the "other directions" that I hinted at, which **Gordon** would take, as compensation for the end of the Moon program when he quit NASA? What could possibly rate in comparison to flying to the Moon? Easy. He became VP of a National Football League Team. Dick Gordon managed the New Orleans Saints from 1972 to 1977. He leaves us with a cameo appearance in a DVD.[12.5]

We read in Chapter 5 how **Fred Haise** (Figure 6.21) had a big part in getting

Figure 6.21. **Fred Haise** at a Tucson, Arizona, space convention, June 2017.
Credit: Author's collection

the Shuttle ready for space flight. He had originally been hoping to fly again to the Moon after the Apollo 13 disaster had taken away his chance, as the primary LEM specialist,[13.2] of getting to the surface.[13.3] **Harrison Schmitt** said that:

"Fred Haise knew more about the lunar module than anybody alive."[19.10]

Indeed, he had been the back-up LMP for **Buzz Aldrin** in Apollo 11. But Apollo 19, where he had the prime crew assignment, had been canceled. He resigned from NASA in June 1979, because he realized that he was no longer as fit as he had been, and his eyes were getting weak.[13.5] Some of his Apollo astronaut colleagues had already been a decade in the "world beyond NASA" before **Haise** ventured forth. But he did not stray too far from his space roots, becoming a VP, and later the President, of a Grumman subsidiary—Grumman Technical Services—with maintenance contracts for the Shuttle and ISS work.[13.2] He came back to the fold (or at least the NASM in Washington, DC), whenever there was a celebration of the safe return of the Apollo 13 crew.

That was where I met him in April, 2010, "being anonymous," walking around the museum where I was a volunteer docent, a few hours before he gave his lecture in which he pointed out that:

"the procedures for saving the crew had been developed in hours instead of the normal process taking weeks—and furthermore, there were no errors!"[13.3]

Haise can clearly see the rationale for an enhanced space exploration agenda:

"The space program is the means, the mechanism, to establish the human race elsewhere ... I don't see any clear-cut thing evolving that's going to say 'here's the funding, and we're going to Mars'."[13.2]

And, finally, the clincher:

"The Creator gave us the capability to do this."[13.2]

I also talked with him in Tucson in 2017 about the Google Lunar XPRIZE (GLXP, see Appendix D), and this new way of returning to the Moon motivated by a prize purse, and he expressed much interest.

"How will you split the prize money if two teams land at the same time?"

"That's why we GLXP judges get paid the big bucks to decide!" I replied. I could tell by his wry smile that he appreciated we were all unpaid volunteer judges! Then, he finished off his thought:

"I sure hope that one of the teams at least manages to land!"

I could understand why that was so important, since he had made that long and dangerous journey on Apollo 13, and yet was never able to make the landing, due to the mid-course explosion. As this book is being written, we are still eagerly waiting to see how the GLXP competition will end.

Jim Irwin (Figure 6.22), was also a firm believer in the Almighty—maybe in part because he had survived a plane crash in 1961, when:

"Doctors said I would probably never fly again."[14.2]

Yet he went on to walk on the Moon. He had developed an irregular heartbeat during his Moon mission, and struggled with health problems after his return, dying of a heart attack on August 8, 1991, aged 61. **Irwin** was another one of the four Moon travelers whom I never met, and whose perspective on such things as space tourism I would therefore never be able to assess. He had been the LMP on Apollo 15, and in concert with the other crewmembers from that mission, became embroiled in the so-called Apollo 15 "stamp scandal" reprimand, which in looking back was a pointless and sad distraction following the crew's mission success. Who cares today how many sets of postal covers—authorized or otherwise—the crew took with them to the Moon?

Irwin's perspective during his mission was very much colored by his religious inclination before he set forth:

"I was getting more aware of spiritual truth before the flight. I wanted to be prepared in every way in case I didn't come back."[14.1]

His Moon mission was his only spaceflight, and he found the experience some-

Figure 6.22. Jim Irwin, in an airport lounge, departs for an expedition to Mount Ararat in search of Noah's Ark in 1989.
Credit: Stars and Stripes

what overwhelming, being able to gaze back at the entire Earth slowly spinning in the surrounding blackness.

"Anyone passing through the Solar System would be attracted to the blue planet ... We, the blue planet, stand out as a beacon to all."[14.2]

Irwin referred to his viewpoint during the trip as a:

"heavenly perspective."[14.2]

"When I came back, my new purpose in life was to share faith, a faith renewed, a faith restored. A faith that came alive on another world."[14.1]

"We are all crew on Spacecraft Earth that is hurtling through space to an unknown destination ... to reach [that destination], we must take care of our spacecraft and each other. We all have new responsibilities since we have shared the perspective."[14.2]

Irwin echoed the sentiments of other returning Moon travelers when he noted the heightened awareness of things terrestrial:

"When I came back after twelve days on the flight to the Moon, I could appreci-ate little things—like being able to eat with a spoon, to be able to lie down in a bed and stay in that position, to be able to smell things, to appreciate the sense of the Earth, to really hear sounds."[14.4]

After the obligatory White House visit, and the world tour, **Irwin** resigned from NASA and the USAF in July, 1972, and created his own Christian ministry in Colorado Springs, the High Flight Foundation, to which he devoted the rest of his life. But his days as an explorer were not over. Over the period from 1982 thru 1989, he decided to use the space perspective and remote-sensing technol-ogy, together with religious scripts, to find the resting place of the Biblical Noah's Ark. He traveled to India and eastern Europe, and homed in on modern day Turkey's Mount Ararat as the likely location, and he identified some land formations, disclosed on the map he carries in Figure 6.22, as the location of the Ark's remains. But his failing health, combined with terrorist activity on the mountain, prevented him from concluding his search of the anomalous land form, which subsequently turned out to be a natural ice cave. We do, however, have the book he wrote to describe his experiences and his transformation,[14.3] and the role that he believed that he had been singled out to play. The book ends with an epilogue in which Irwin struggles with his recent heart attack, and intimations of mortality. The book's title conveys his strong spiritual focus: *To Rule the Night.*

Jim Lovell (Figure 6.23), had been, he asserts:

"interested in astronomy before the Glenn's and Shepard's of the world could spell 'rocket,' and in 1952 I wrote a term paper at Annapolis on the development of the liquid fuel rocket."[15.7]

It is hardly surprising, therefore, that he would not be satisfied by merely orbit-ing the Moon, as he had done on Apollo 8, but that he would want to go back, and this time would plan to land on its surface. The story of that failed second trip, Apollo 13, was made into a successful movie, based on his book of the experience: *Lost Moon.*[15.9] Figure 6.23 records the moment when the actor playing **Lovell** in the movie is greeted by the real **Lovell**. It is hardly surprising that the Moon travelers sometimes had a hard time of it facing reality on their return to Earth; there could be times when the boundaries became blurred. **Lovell** had sold the movie rights even before the book was written,[15.7] so we know that his post-Apollo years would be a massive transformation from the days when he had washed dishes and bused tables at Gammon's Restaurant in Madison, Wisconsin, back in 1946.

In 2009, **Lovell** would have his own highly successful restaurant, "**Lovells** of Lake Forest," in Illinois, with his son James as chef. His book and movie came out in 1991 and 1995, and they led to a career on the talk circuit. After retiring from NASA and the US Navy in March, 1973, and before the success of his book, **Lovell** had tried various ways to occupy himself following his two Moon journeys. He had tried unsuccessfully to obtain one of the coveted Coors Beer

Figure 6.23. A scene from the movie *Apollo 13* released in 1995, with Tom Hanks (on the right—playing **Lovell**) meeting the real **Jim Lovell** (on the left—playing the cameo role of the Commander of the recovery carrier fleet), when the crew is greeted on their safe return following the catastrophic explosion on their way to the Moon. Compare with the real thing, back in 1970, in Figure 4.35.
Credit: Screenshot

distributorships. Remaining initially in the Houston neighborhood, he first of all spent 4 years in the boat business, as President/CEO of Bay-Houston Towing Company,[15.7] then discovered the telephone industry as President of Fisk Telephone Systems, from 1977 to 1981. He moved to Chicago, his new family base, in 1981, and was Executive Vice President of Centel there until he retired, with the success of his book and movie in 1991.[15.7] He continues to serve as a director on various Boards (e.g., Federal Signal Corporation), and is a motivational speaker on the lecture circuit for such organizations as The National Space Society and the Space Foundation.[15.7]

Lovell, having survived the near-disaster of the Apollo 13 mission, is very content with his life:

"I don't worry about crises any longer."[15.7,15.10]

Nowadays, he still enjoys flying and hunting. His take on the future of mankind in space? Well, he is not given to philosophy, and his approach is still to monitor whatever NASA is doing, rather than what is happening in the commercial space sector. He was not impressed with all the years devoted to building and operating the ISS:

"We wasted an awful lot of money, time and effort."[15.7]

He also is an earnest advocate of taking more risk in space endeavors.[15.2] And so **Lovell** wrote to me in 2016,[15.6] exhorting me to keep working towards more exploration and maybe settlement.

"Mars in maybe 20 to 25 years"

was what he had earlier predicted.[15.7]

That would have put the Mars landing between 2019 and 2024.

As we saw in Figure 4.9, nobody who went to the Moon is ever really free of that experience, and he continues to serve NASA by giving his public talks, at NASM, the Kennedy Space Center, and elsewhere. **Lovell** has a wonderfully dry sense of humor, and his take on history is always a delight to hear again.

"If you buy that Glenn's second flight into space is about research into aging ... then, I've got a Bridge in Brooklyn to sell you!,"

he said to a February 2002 gathering at KSC, which included the former Senator, and reflecting 40 years since Friendship Seven.

You can watch some great cameos provided by **Lovell** on two DVDs.[15.11,15.12]

It was not until 1985 that **Ken Mattingly** (Figure 6.24) finally hung up his space boots, as we saw in Chapter 5. In that year, he had been the last of the former Moon travelers to be still carrying out space flight. By this time, he was nearing 50. He took over responsibility for the procurement side of the US Navy's electronic systems (i.e., satellite communications in the Navy) as Commander, US Navy Electronics Systems Command.[16.5] Unlike many of his former colleagues, however, he became an early supporter of the "new space" movement, which advocated for the commercialization of space exploration and exploitation.[16.7]

I saw him in October 1996, with **Pete Conrad**, at a conference of the Space Frontier Foundation in Los Angeles. Both of them were trying to support new developments, including the notion of space tourism. **Mattingly** said:

"Someday in the future this tourism stuff is going to be a blockbuster."[16.6]

He had always hoped that, someday, he might himself go on a Mars mission, although he realized that:

"we are now risk averse."[16.2,16.5]

He had to be satisfied with more modest goals for the remainder of his professional life. When **Conrad** had been developing the McDonnel Douglas DC-X, **Mattingly** was pushing the Lockheed Martin X-33. Both craft were aiming at re-usable low-cost access to space; neither one succeeded, although **Pete** managed to get in a few test flights before cancelation. I saw something of the way in which **T.K.** cared for his team members when, in August 2001, he was telephoning around in person to try and place members of his crew who were being laid off due to lack of funding. He was chasing up on a job announcement in my organization for an analyst position as a potential solution for one

Figure 6.24. **T.K. (Ken) Mattingly** at a conference in Los Angeles on October 19, 1996, just a decade after his last spaceflight, where he had taken part in panels discussing commercial re-usable space planes.
Credit: Author's collection

of his staff. Even when told that this would not fit, he still came back, suggesting even a short-term appointment. That's the kind of boss I would have liked to have had!

Mattingly ended his NASA career remaining connected with the space program, with positions as an executive with a Grumman space station contractor[16.5] and at General Dynamics in San Diego as program chief of the ever-lasting Atlas vehicle. He has been remiss amongst the Moon travelers, in not writing a book, so we don't know his detailed thoughts about current developments, but[1.2] Chaikin reports him as saying:

"We are in danger of forgetting how to explore space. If you do not build things, you don't know how to build things."

We can assume, therefore, that he ended his professional days being frustrated with the slow pace of change, at least when compared with the heady pace during the 1960s.

With his death on February 4, 2016, aged 85, almost exactly 45 years after his Moon trip, **Edgar Mitchell** (Figure 6.25), signaled a sad fact. Now there

Figure 6.25. Edgar Mitchell continuing his search for a unifying theory of mind, matter, and energy at a conference in Washington, DC, on July, 2007.
Credit: Author's collection

were no survivors left of the Apollo 14 mission. **Roosa** had died in 1994, aged only 61; **Shepard** had died at age 74, in 1998. **Ed Mitchell**, the LMP for Apollo 14 always regarded himself as an explorer:

"Exploration is in my blood,"[17.10] *... You can't explore without risk,"*

noting that he came from

"a family of pioneers who settled West,"

his father being a cattle rancher.

He had his first flight from a barnstormer in a Jenny when he was only 4, soloed in a Piper Cub when 14,[17.7] and was a licensed pilot by the time he was 16.[17.1] He grew up in New Mexico near Goddard's ranch, and so saw some of those early rocket experiments.[17.7] The book he wrote on his return from the Moon is even titled *The Way of the Explorer*, but it is only partly about exploring the Moon, where in fact he had a lot of trouble navigating around on the flanks of Cone Crater:

"Our positions are all in doubt now."[17.10]

Its main focus was instead on recording, and investigating, a transformational experience *"a personal epiphany,"*[17.10] he had on his return journey to Earth.

Mitchell would spend the rest of his days pursuing this other exploration. He described the initiating experience,[17.12] as:

"an ecstatic feeling of oneness with the universe, and oneness with nature ... I suddenly felt very protective of Earth."[17.7]

This was not for **Mitchell** a religious experience; he was agnostic. **Mitchell** was a deep thinker, very talented (he played the violin, viola, piano, and trombone, for instance), and was generous and encouraging to others (he gave me a lot of his time to explain his theories[17.5,17.6,17.8,17.9]), and wrote a Foreword for a book in which I was a co-author.[17.4] He put a great deal of emphasis on the need for scientists to be open-minded in order to be able to recognize and assess possible new data and insights. He therefore relied a lot on intuition.[17.10] He was not a romantic at all, however, and stated:

"I don't recall ever reading science fiction. I was trying to make fact, not fiction."[17.10]

His life's work after returning from the Moon was:

"all about discovering ourselves and our place in the cosmos."[17.10]

Mitchell left NASA, and the US Navy in 1972,[17.1], when he was still only 42. A year later, he founded the Institute of Noetic Sciences to study consciousness and related phenomena. This was an enormous personal risk for him, since it placed him outside of the comforts of conventional scientific thinking, and exposed him to possible ridicule. But he saw himself as a scientist trying to establish a new discipline, which needed to be given a structure, and a means of experimentation and measurement. There is no question that **Ed Mitchell**'s personal Moon flight experience was the most dramatically changing of those of all the 24 Moon travelers.

"We went to the Moon as technicians and returned as humanitarians."[17.3]

He left behind his Institute, his book, and his DVDs[17.11,17.13,17.14]—also the Association of Space Explorers, of which he was a co-founder in 1984. Peace, love, saving the planet, hope, tolerance,[17.8] were the messages he delivered on the lecture circuit. He could see the need for space colonization but wanted global civilization first.[17.8] He struggled with trying to understand the role of the human mind in the grand scheme of things, and on a more personal level with how a scientist/engineer, such as himself, could integrate more closely with society in general.

"We did not know how to talk about our feelings ... I didn't know what feelings were. People used to ask 'what did it feel like to be on the Moon?'—I didn't know what it felt like. I could tell them what I did, and what I thought, but not how I felt."[17.1]

I put it to him in 1997,[17.6] that his "epiphany" during his 3-day return from the Moon was simply an adrenalin high, as a result of a supreme challenge satisfactorily met, and that his open-mindedness was laying himself out there as a target for quacks and charlatans to abuse, but he remained insistent. I cannot now imagine how **Mitchell** got on with his boss **Al Shepard**. Except that **Shepard** recognized in him his straight shooting, and his extreme scientific and engineering competence. They could not possibly have been more different as people. **Mitchell** could see the eventual destiny of mankind *Beyond Earth*, to quote the title of the book for which he wrote the foreword.[17.4] He had written to me that

"space tourism will bring great benefits,"[17.5]

and was certainly thinking of the "Overview Effect," and his hope that *"a few notables,"* if they could only see the Earth from space, would moderate their destructive behaviors. Maybe, just maybe, **Ed** could see further than the rest of us. Certainly, this kind and gentle man cannot be faulted on his courage and steadfastness, his message for peace, and for protecting the home planet.

Apollo 14 had gone to the Moon with the lowest cumulative crew spaceflight experience of any Moon mission that had gone before or for that matter subsequently. Two of its crew were complete "rookies." **Ed Mitchell** had made only the one spaceflight. The other rookie in that Apollo 14 mission was **Stu Roosa** (Figure 6.26), who also did not fly again. Of course, **Al Shepard**, the mission Commander, was no rookie—he had a cumulative 15 minutes of spaceflight when he led the trip. **Stu** retired from the USAF and NASA in 1976, after having been on the "dead-end" back-up crew of Apollo 17. That was one mission which was certainly never going to allow the back-up crew to fly, for sure—as was tellingly pointed out earlier (Figure 4.60) by **Ron Evans'** widow.

Figure 6.26. Stu Roosa spreading the word about saving the planet via his Moon seeds.
Credit: Forest History Society

Roosa had been personally impacted by the death of friends and co-workers.
Several had died in crashes at Edwards. He had witnessed the death of a fire-
fighter during his "smokejumper" days back in 1951. He nearly crashed himself
when he flew a plane with an inoperable fuel pump switch. And he was in the
blockhouse with Deke Slayton, acting as CAPCOM, when he listened to the
Apollo 1 crew burning to death. Whether these events somehow influenced his
thinking or not is unclear, however, he was religious (Catholic), and saw his
Moon mission from that perspective. He believed in Divine Intervention.[18.2]
He was also a close friend of **Charlie Duke**, who shared the religious perspective. He died in 1994 at only 61, and I never had the opportunity to talk with
him directly, so I only know his views on space exploration from what has been

recorded by others, and I do not know at all what he would have thought about space tourism, since he died before it was even on the table as a possibility. Incidentally, this particular Apollo crew did not "hang out" together much after they returned from the Moon. **Ed Mitchell** reported, at the time of **Roosa**'s death, that the last time they had been together was 6 years earlier.

This is what **Roosa** has said on the record about exploration:

"Exploration is why we're no longer huddled in caves. This spirit which took us to the Moon is the same spirit that moved our forefathers West across the country. And as they carried the flag west, why, we carried it on to the Moon."[18.1]

After leaving NASA, **Roosa** traveled a bit, visiting Nepal and Egypt. And he initially took a job based in Athens, Greece, as the VP of a Middle East development company, US Industries, Inc. That turned out not to be a satisfactory follow-on to his Moon trip experiences, and after a year, he opted instead to work in real estate as VP of planning at Charles Kenneth Campbell Investments (which was owned by one of **Roosa**'s Big Game hunting buddies), where he remained for 4 years.[18.2] Finally, he took the well-trodden Coors route, and he owned a Coors beer distributorship, as President of Gulf Coast Coors from 1981 for 13 years until his death. So much for his professional activities after his return from the Moon. But they did not define him. Despite his earlier protestations to the contrary, his Moon trip had changed him. For one thing, he re-assessed his Cold War activities after visiting Moscow in 1991.

"After seeing this, I sure would have hated to have blown it up."[18.1]

He became comfortable in his skin:

"I think, from within, it gives you some confidence. I think if you can handle it right you can decide that, hey, I have to stop trying to climb for the top of the mountain, because I've already been there."[18.1]

This is how **Stu** used his temporary fame to try to make a difference:

"We addressed the General Assembly at the United Nations. I made the comment that, if everybody in this room had just had the view of Earth that I had, then the discussions would go a lot smoother."[18.1]

But his real ongoing legacy is seeds. Yes, in commemoration of his "smokejumper" days, putting out fires in the national forests, **Roosa** had taken a bunch of tree seeds on his Moon voyage. Why not? They were extremely lightweight items which would turn into hundred-foot trees! So, on his return he distributed seedlings (Figure 6.26). They have now become well-established trees, there are even second generations, and there may even be one near you, with an associated plaque declaring its provenance. They are all over the US, and some are abroad. Just Google "Moon trees"! Even if you never met him, you can shelter under one of his trees, of which there may well now be thousands. They are a living connection to that first golden age of space exploration. So, in a very real sense, **Roosa** had paid tribute to his forestry roots, while contributing in his unique way to saving the planet, to which he had referred in his UN speech.

Harrison "Jack" Schmitt (Figure 6.27), is an effective politician and geologist. He said in March 2003 at a lecture in Washington, DC,[19.6]

"I hope I can still call myself a field geologist."

He had been the last of the 12 lunar explorers to set foot on the Moon, and of course was the most qualified to study its surface when he got there. He had worked hard at making sure that the Apollo missions brought back useful scientific results (which of course had not been their initial motivation). **Schmitt** came up with the idea of the Science Support Room at Mission Control,[19.10] worked on the Apollo 11 rock samples, and authored the first scientific paper on their characteristics. He remained involved with NASA science until resigning, in 1976, when he was elected as a Republican US Senator for New Mexico for a 6-year term. There he honed his skills as an effective Committee chair,[19.11] which would later be brought to the aid of the space program as he chaired the NASA Advisory Council (Figure 6.27) in 2007 and 2008. I observed from the public seats as he expertly handled the diverse subject matter, and interjected his own clear perspectives, while displaying an upbeat, action oriented, and concerned leadership to the proceedings. And a playful sense of humor was never very far from the surface.[19.7] Around that time, he brought out his book *Return to the Moon*,[19.12] in which he makes a concerted attempt to evaluate the Moon as a source of energy resources for the Earth, particularly emphasizing the relative abundance of Helium-3, which could be a clean fuel for fusion rectors, in its surface material, or regolith.

Schmitt had been Adjunct Professor of Engineering at the University of Wisconsin after he completed his term as Senator. He was an early supporter of the "new space" agenda, and he lent his support in 1996 to the Space Frontier Conference in Los Angeles. In 1997, two decades ago, I stood in line with him[19.8] in the speakers' registration line for an Albuquerque, New Mexico, conference on Engineering, Construction and Operations in Space. He was pitching his Helium-3 plans; I was talking about using the dying Soviet space station *Mir* as a potential space tourism hotel. He was immediately practical:

"The problem would be insurability ... it would need considerable refurbishment."

Schmitt has been a constant presence in support of commercial space initiatives in his home state of New Mexico. I saw him at Spaceport America in October, 2010, when the runway was opened and dedicated in the name of his retiring State Governor Richardson.

Schmitt knows full well that the old-style Apollo model is not going to cut it any more. He is a realist as a rare blend of scientist and politician:

"Those of us who are interested in going into deep space have to concentrate on finding a commercial reason to do it."[19.10]

Figure 6.27. Jack Schmitt in his role as Chairman of the NASA Advisory Council, February, 2007.
Credit: NASA

"I don't think the government's ever going to go back and tap these energy resources [on the Moon]. I think it's going to be a private-sector initiative."[19.11]

"The only way that science is going to go back to the Moon is Piggyback."[19.10]

"Space is vast, packed with action[19.4] *... it challenges us to use extraordinary technology. Early explorers of the sky not only went into space, and became the eyes and minds of billions of others. They began the long process of transplanting civilization into space. With the conclusion of the Apollo 17 mission, humankind reached 'the end of the beginning' of its movement into the universe ... We've changed our evolutionary status in the universe*[19.6] *... We know we can now live on the Moon and Mars, should we choose to do so. We can settle there. The resources of the Moon can support indefinitely and independently settlements in space by human beings."*

Schmitt remains a committed and engaging public speaker on space, science, technology, and the environment, and furthermore he has supported the cause of space tourism,[19.9] the need to be less risk-averse,[19.3] and space architecture ideas such as Gateway Earth (Appendix E)—which uses space tourism revenues to enable further space exploration. In that context, he provided me with valuable insights about the viability of a hotel for space tourists in geostationary orbit.[19.9] **Schmitt** lends his experienced voice to those providing cameo appearances in two DVDs.[19.5,19.13]

Dave Scott (Figure 6.28), had early experience of Europe when around 1960 he served as an Air Force pilot stationed in the Netherlands. And Europe would feature prominently in his post-Apollo years. His Apollo 15 flight, with the first use of the Lunar Rover, and the exploration of Hadley Rille, had been a triumph for science. After dealing with the aftermath of the Apollo 15 "stamp incident," when there was an in-retrospect trivial disagreement over the number of postal covers the crew had taken to the Moon, he served in NASA management roles including involvement in the ASTP, where he was not only on the back-up crew, but led a NASA delegation to Moscow in 1973 to sort out technical details of the upcoming joint US–Soviet mission. I always recall this period with fascination. Here were opposing military officers, in the middle of the Cold War, working together and trusting each other with their data and their lives. The impact of that experience would have to be sobering. This experience would eventually (in 2004) prove more widely fruitful when he jointly authored a book[20.3] with the former soviet cosmonaut Alexei Leonov, with whom he had worked on that assignment.

I met him, with Leonov, in August 2005, as part of his book tour,[20.2] and briefly discussed space tourism with him. He had retired from the USAF in 1975 to become Director of NASA's Flight Research Center at Edwards until 1978, when he entered the private sector.

"When I landed on the Moon, and came back from the Moon, I was 39 years old ... My career had been finished. Now go find a new career."[20.1]

Figure 6.28. Dave Scott at a space conference in June 2011.
Credit: Collectspace.com

He missed the clarity of military directions, and of the Apollo challenge. He said:

"It was probably the clearest definition of an objective that mankind has ever experienced—"Man, Moon, 1970". How could it be any clearer than that?"[20.1]

What he did was establish various consulting businesses, such as **Scott** Science and Technology, mainly involving commercial applications of space technology, such as opto-electronic sensing, and for a period moved his base of operations to the UK. He also served 4 years on the COMSTAC committee of the FAA, one of my own later committee assignments, to advise on commercial space-flight matters. **Scott** had fully appreciated the Apollo experience, and wanted to share it and motivate people to press on further, and was willing to try any means to do so. He had taken a small piece of art—*Fallen Astronaut*—by Belgian sculptor Van Hoeydonck, to leave on the Moon representing humani-ty's aspirations. And he spread the vision later during his private sector years through consulting for TV and movies about space (including the movies *Apollo 13* and *Magnificent Desolation*).

Listen to how he described why it is not enough to just send robotic craft into space:

"Oh, but the reason you send people is so clear. You can't explore without the perception, judgement and awareness and the intuitive nature of man. Man is going to explore the universe, and pioneer and settle the universe. There's no ques-tion; it's just when. You don't have to do it this year or next year, but it'll get done."[20.1]

He provides us with a perhaps necessary reminder of what had been achieved:

"Look at the system we had, and how well it worked. Not a single LM failure. Not a single backpack failure. Not a single rover failure ... that's magnificent!"[20.1]

We so easily forget that a failure in any of the thousands of parts would have been fatal for the crew. He has no doubt about where we are going, either:

"Apollo had a reason. It taught us how to go in space and set up our first outpost. Now we go to Mars."[20.1]

Well, that had not at all been the reason at the outset, but that's what it became, until the money ran out. He discussed the Gateway Earth approach (Appendix E) with me in Tucson in June, 2017, and was much taken with its possibilities: He commented:

"What a great idea!"

He still works as a commercial space consultant, now based back in the US, currently as President of the Baron Company. We have **Scott** to thank directly for his work in putting the space story so technically accurately onto our movie screens and DVDs. You can watch him in person do a cameo role in one DVD.[20.4] And Ron Howard and Tom Hanks indirectly owe him for some of their Oscars!

By choosing to relate these post-Apollo stories alphabetically, we only now come to **Alan Shepard** (Figure 6.29), who is more usually the *first* in any space commentary, because of his mission in *Freedom Seven* as the first American in space. **Shepard** died July 21, 1998, aged 74, just before his old Mercury buddy

Figure 6.29. A relaxed **Alan Shepard** showing the model of the golf six-iron he used on the Moon during his February 6, 1996 visit to the Far Hills golf club in New Jersey.
Credit: NJ.com

John Glenn got his second flight into space taking a back seat on a Shuttle ride. **Shepard**'s own second flight had been to the Moon on Apollo 14. He had spent a frustrating period removed from flight status due to his balance problem, while everybody else was getting their rides. And he was, with Deke Slayton, in charge of that process. And he could be grumpy.

But after he returned from the Moon, he was a changed man. He had been hyper-competitive, and now he had achieved the pinnacle. He was the only one of the original Mercury Seven astronauts to reach the Moon—and even walk on its surface. According to **Charlie Duke**:

"Al seemed easier to get along with after Apollo 14. He was more like 'one of the guys'—not aloof. He could still be a stern boss, but I thought he was more approachable."

Shepard himself recorded later that his father was at last proud of his son's achievement. That was perhaps the biggest change of all. Being the first American in space had just not been good enough. **Shepard** stayed around until his buddy Deke finally got his spaceflight, with ASTP, then quit the Astronaut Office and the US Navy on August 1, 1974.

"Life has to go on ... You've got to define new challenges."[21.1]

He wasn't much of a philosopher, but he summed up his views thus:

"I've always been a believer in pushing out the frontier—it was kind of nice that the Moon got in the way."[21.1]

He also registered

"an overwhelming feeling in seeing the beauty of the Planet [Earth] on the one hand, but the fragility of it on the other."[21.3]

Dave Scott had this conversation with **Shepard** when **Al** came back to the astronaut office after being on the Moon:

"I said 'Hey, Al, what's it really like?' And all he had to do was look me in the eye, and I could read it. And he said 'It's spectacular,' and that's all I needed to know."[20.1]

By the time he left NASA, **Shepard** was already a rich man. During his remaining two decades, he consolidated this by being on the board of dozens of companies, and made money from real estate investing (Marathon Construction). And he even had one of the highly sought-after Coors beer distributorships. His joint book *Moonshot*, written with Deke,[21.5] was a bittersweet achievement, since Deke died just before it was published in 1994. He remained as President of the Mercury Seven Foundation, which he had set up at the beginning of the space program, but he was not otherwise active in space matters.

After the Moon missions, golf was his overriding passion, and he moved his home to Pebble Beach, just by the famous golf club in California, so that he could enjoy it. **Shepard** had, you will recall, hit a golf ball on the Moon, and in Figure 6.29 you can see a replica of how he had managed to get the club aboard the Lunar Lander. It was assembled from parts of his lunar equipment, and he had simply brought the club head to be added when his Moon duties were over. He was always careful to state:

"The golf shots were at no expense to the taxpayer."[21.3]

Shepard would turn up, though, for a space event, if it was being organized for some charity by his Mercury buddy Wally Schirra down the coast in San Diego. So that is where I met him 3 years before he died, around about the time of the function recorded in Figure 6.29, at another charity golf event— "Shoot for the Moon." This was an invitational golf tournament and auction dinner in support of the San Diego Aerospace Museum, where he handed out prizes—of golf clubs—to the winners. He was very contented with his life. He talked briefly about space, but mainly about golf. After he died, and his wife Louise died very shortly afterwards, their ashes were scattered from a naval helicopter hovering over the Pacific Ocean coast, just off the edge of his beloved Pebble Beach fairway.

   Tom Stafford (Figure 6.30), by contrast, has remained connected with space and military policy ever since his NASA duties ended. The biggest change he experienced was not his going to the Moon, but going to Moscow. He makes his former mindset very clear:

Figure 6.30. Tom Stafford (*left*) with his old Soviet counterpart and lifelong friend Alexei Leonov, at a 50-year Yuri Gagarin ceremony in Moscow, April, 2011. Leonov continues his support of space tourism and has with him Helen Sharman, who was an early private space traveler, flying 20 years earlier in May 1991 to the Soviet space station *Mir*, having been trained under Leonov's tutelage.
Credit: The Independent

"I was originally a Cold Warrior—I wanted to go to Korea and shoot down MiGs, and kill Commies."[22.6]

But during the ASTP activities, he developed a deep friendship for his former Soviet fighter pilot adversary, Alexei Leonov.[22.9] In the process of developing cooperation, they had each learned the other's language. And had drunk many vodkas, one suspects. **Stafford** picked up on one Russian tradition when he arranged for the planting of celebratory trees outside the National Air and Space Museum in Washington, DC, to commemorate the friendship of the US–USSR crews of ASTP, the mission which had effectively heralded the coming of the end of the Cold War. The trees are still there, if you can find someone, like an old volunteer docent, to point them out to you.

Although he retired from the USAF in November, 1979, **Stafford** is still very much a military man, not by any means a romantic. For instance, he has said:

"The finest hour, in my viewpoint, of the space program was getting Apollo 13 back, not the first lunar landing."[22.1]

Stafford appears regularly on Capitol Hill responding to Congressional enquiries about space, mainly with reference to the current mainline NASA program at the time, and has not been persuaded by me, though I tried,[22.8] to embrace the commercial space approach. In June 1991 he chaired the study "America at the Threshold," which however did not lead to any marked changes in US space policy. He has assembled, by some remarkable process, an amazing collection of aerospace artifacts, at his own Stafford Space Museum in his native Oklahoma. And he has his book *We have Capture*,[22.10] the title of which records the statement made at the moment that the US and Soviet parts of the ASTP mission rendezvoused and docked in space. **Tom Stafford**'s post-NASA professional era was more or less split in two. Initially his work was USAF related—in August 1975 he became Commander of the USAF Flight Training Center at Edwards. Then he gravitated to DC, where he became Deputy Chief of Staff at USAF HQ—working on testing the first stealth experimental aircraft.[22.7] From 1980 onwards he became a consultant on defense technology (his firm being **Stafford**, Burke and Hecker), and never strayed very far outside the Washington Beltway (unless dealing with his museum activities). Even though one can read it all in the books, including his own, there is still a freshness when the information is coming from someone who actually flew to the Moon.[22.4] For example, at a "Moon, Mars, and Beyond" Aldridge Commission hearing at the Department of Commerce in Washington DC, in February, 2004,[22.8] **Stafford** was giving evidence about going back to the Moon or further, and made a plea for improved space suits:

"We need new lightweight pressure suits, with better gloves—it's like doing a penmanship contest wearing boxing gloves ..."

On program management, he offered this:

"Hire good people and trust them. Avoid the pitfalls of interminable studies."[22.8]

Still good advice, all these years later. He regrets the absence of the clarity of earlier eras:[22.2]

"What is our challenge today? Are the first astronauts to Mars in our grade schools now?"

I expect we shall be continuing to hear from **Stafford** any time that space policy is up for discussion on the Hill. But expect that cooperation will be a big part of the agenda. He talked about the trees he had planted at the NASM after his ASTP flight:

"Yes, it's an idea I learned from the Russians. Somebody said we should keep the seeds. The future is cooperation. We knew we would need them some day."[22.8]

And that cooperation with the Russians continues to this day with the ISS, and indeed as I write the former Soviet Soyuz is still the only craft getting US astronauts into space, and bringing them back to Earth. That will hopefully soon change, when the commercial approach that **Stafford** has been perhaps underplaying, kicks into gear and provides a relatively cheap taxi service from US spacecraft into low orbit from US spaceports again. His buddy Leonov's Soyuz has indeed done sterling work for the space program in general, and I might even say, looking at Figure 6.30, for the birth of space tourism (all 10 of the original orbital space tourism flights used Soyuz). But Soyuz, like some of the veterans of the space program, has been flying since 1967. That's 50 years. Maybe it is time for retirement, and a new generation.

Jack Swigert (Figure 6.31), is another of those Apollo-era Moon travelers who died too soon. He died in 1982 of bone cancer, and I never met him. Nor did he write his own book. Of course, the Hollywood version of his part in the Apollo 13 story is recorded in the book and associated movie.[23.1] We know of him therefore mainly from his former colleagues. **Fred Haise** said:[13.5]

"Jack was a real easy going guy."

He was a Command Module Pilot with considerable experience before stepping into his Apollo 13 seat at short notice, but he knew little about the Lunar Module until he was, perforce, to fly in it all the way home after the onboard explosion. As **Fred Haise** pointed out,

"Jack's first time inside the LEM was in flight!"[13.2]

We can see from Figure 6.31 that **Swigert** decided to emulate Glenn and **Schmitt**, and enter politics. He was a Reagan Republican, running on a technology platform. **Swigert** knew of his cancer diagnosis as he continued his campaign:

"I was privileged to be one of the few who viewed our earth from the Moon, and that vision taught me that technology and commitment can overcome any challenge."

And, from Figure 6.9, we see that he succeeded. He became a Colorado

Figure 6.31. Jack Swigert runs for congress as a Republican in 1982.
Credit: Pinimg.com

Senator, although he died before being able to take office. After his mission on Apollo 13 in 1970, **Swigert** had remained with NASA for several years, including supporting the ASTP mission, and early work on the Shuttle, according to **Mattingly**,[16.5] but he never flew into space again. His involvement in politics first took shape when, in April, 1973, he became the Executive Director of the Committee on Science and Technology, in the US House of Representatives. **Swigert** eventually resigned from NASA, in August, 1977, to begin his personal political quest. He supported his activities in this period initially by operating as a consultant in Virginia, and later by being VP of BDM Corporation of Golden, Colorado, and of course through the normal political fund-raising letters, in his case linked with autographed space memorabilia gifts:

"Enclosed are your [philately] covers signed as requested. Now, I need your assistance ... to help me reach those people I will not meet personally,"

reads one in the author's collection.

Swigert tried unsuccessfully for a US Senate seat in 1978, but was elected in November 1982 from Colorado to the US House of Representatives, dying a month later at age 51, scarcely a decade after his fateful Moon mission. What did he think of the prospects of space tourism? What about his views on long-term space exploration challenges? We shall never know. He died too soon, and without sufficient record. **Swigert** never married, and so there are not even any children to perpetuate his tale. But the wonderful statue at Denver International Airport (Figure 6.31) continues to provide inspiration to air travelers, such as myself, as we hustle between gates. And it's a kind of bonus that there is a replica on Capitol Hill to serve the same function, but in this case, maybe a greater percentage of lawmakers receive the very necessary motivational message.

Al Worden (Figure 6.32) now spends a lot of his time on the talk circuit, where he has developed a reputation for telling a good tale. I met with him in Oxford, England, in 2015 at an event at the historic Bodleian Library Divinity School, and can confirm that perspective. He gave a good account of his Apollo 15 mission, and the EVA that he undertook between the Moon and the Earth during the return flight—

"The first Lunar EVA,"

which Worden insistently and correctly calls it. Showing a somewhat blurry image on the big screen, he humorously pointed out:

"Jim and Dave had thousands of photos of themselves on the surface. For my trans-lunar EVA, they took only this one! Showing my butt! I think it was a conspiracy!"[24.5]

Worden had always stressed that Apollo 15 was a scientifically focused mission,[24.4] and proudly underlined the important role he played managing the Command Module's SIM-bay (Scientific Instrument Module) experiments, which included an X-Ray scanner, throughout the mission. He claimed that he did not think much of:

"the other two guys. What were they doing? Just picking up rocks to provide ground truth for my experiments."[24.6]

Worden gave a wonderful description[24.6] of what he saw when outside the spacecraft en route somewhere between the Moon and Earth:

"What I remember from my EVA was a blinding light of stars. Just so many stars. You realize that there just have to be many of them with Earth-like planets."

As the years have progressed Al Worden has mellowed. He had been difficult, or maybe we can simply say complicated, earlier, fretting for decades over the fallout from the "stamp scandal" which had erupted at the end of his Moon mission. This is even captured in the title of his second book: *Falling to Earth*. At a book event in 2012 he said:

"It took me 40 years to write this book, because I didn't want to write it."[24.6]

In the book he describes[24.7] how he eventually successfully sued the US government, his former employer as a NASA astronaut, to have his "Moon

Figure 6.32. Al Worden gives talks, in 2011, about his book *Falling to Earth*. The title underlines the post-Apollo struggle that all the Moon travelers had to take on.
Credit: NASM

covers"—postal philatelic envelopes taken to the Moon, and the subject of a post-flight investigation—returned to him. I suspect that writing the second book allowed him to get things off his chest, and allowed him to put former perceived injustices behind him, and to move forward. He found that, much later in life, there are many folks still out there wanting to hear the stories told, one more time, of when men went to the Moon—but maybe without the controversial or political angles.

My personal favorite, however, is his first book.[24.3] This is a wonderful selection of poetry which **Al Worden** had written describing his astronaut and Moon traveler experiences. I did say he was complicated. All the wonderfully creative philosophical and upbeat material he had written in that book sadly never found much of an audience. "What's wrong with these astronaut guys these days? Artists? Poets? Pshaw!", we can imagine the general public saying, as it was tiring of the Moon missions.

His Moon flight was in July, 1971. What has **Worden** been doing in the nearly 50 years since then, besides churning over the stamp scandal stuff? Until about 1975 he was at NASA Ames as Chief of Systems Study Division, at which time he resigned from NASA. Over the years he has tried various career directions. He was one of the ex-astronauts who considered politics. He reported:[24.4]

"I ran for Congress in Florida, and lost."

Then he moved into energy management consulting, built up various entrepreneurial companies, and sold them.[24.4] In so doing, he also:

"licenced some inventions."[24.4]

In 1993, he became Business Development Executive for BF Goodrich—and Staff Vice President. **Worden** is an Anglophile—he enjoyed his period of test pilot training at Farnborough. So it was in England (Oxford) when I asked him about space tourism, and future directions in space flight. He was not responsive then:

"What is commercial space? We are trying to do it on the cheap, and I think safety will be what suffers. This Obama government has thrown away our leadership in space—and this is important not just for space but for all technological developments,"

and for good measure, he added,

"This Branson stuff is nothing."[24.5]

I have not given up on him completely, though. He is definitely softening up. When I told him that I was a Judge for the Google Lunar XPRIZE (Appendix D), working to bring about low-cost access to the Moon by awarding prizes to teams of young people building hardware in garages, he smiled broadly and said:

"Good for you!"

I am pretty sure that **Al Worden** would like to know that a new generation could contemplate making the journey that he had undertaken all those years ago; even if this time it would not be paid for in quite the same way.

Bringing up the rear of this 24-truck train is the old caboose **John Young** (Figure 6.33). **Young** had the most extensive record of all the astronauts whose flights included the Moon journey. And in his own case, he went there twice. His very first space flight was in Gemini with Gus Grissom back in March 1965; his last was in November 1983 on the Shuttle. He remained as Chief of the Astronaut Office until 1987, at which time he left the US Navy. Between 1987 and 1996, **Young** was the non-flying Special Assistant to the Director of the Johnson Space Flight Center (JSC), with focus on Engineering, Ops, and Safety. From 1996 to 2004 he was the Associate Director for Tech, Ops, and Safety at JSC. Then, after a significant period in senior NASA management, and 32 years after he returned from the Moon for the second time, at the age of 74, he finally closed the door behind him and retired.

In 2012, his biography *Forever Young* appeared.[25.6] And, fortunately, we have great cameo appearances in two DVDs,[25.3,25.4] which capture his laconic style, to enjoy. **John Young** has a wonderful

"wry, low key, sense of humor,"

as reported by his former Gemini co-pilot, **Mike Collins**.[8.6] **Tom Mattingly** reported:

"John may have had a hard time articulating it, but you'd better believe his instincts, because, boy, they are good."[16.5]

Figure 6.33. **John Young** attending a Houston conference in October, 2002. He remained
a NASA employee until 2004.
Credit: Author's collection

I had the good fortune to hear him give two talks, both in the era when he was
still a government employee working for NASA. Although you would not have
believed it from his remarks. Asked about NASA's role *vis-à-vis* promoting
commercial space, he said it had not been done very well, and added:

"It's the law! We should all be in jail!"[25.5]

When I asked him about space tourism, he responded:

*"I think it's a great idea. You know, for years I have wanted to put a module in
the back of the Shuttle, where you could haul 25 or 30 people up there, like in the
back of a bus. Then you open the [Space Shuttle Bay] doors and you have
windows and all there, and you just fill the thing up with barf bags!"*[25.5]

Young recognizes the need for settlement, expressing his concerns over the various potential ways that life on Earth could come to an end, such as mega-volcanoes and asteroids.[25.5] An ESA astronaut, Dr. Gerhard Thiele, who was a fellow speaker at a Naples space conference in October, 2012, described to me an incident when he had been at JSC training for an upcoming Shuttle flight, where the astronauts were discussing the scientific reasons for their flight. **Young** had come in and interrupted the session saying:

"You're all wrong. The reason we are doing this is about the survival of the species ... We have the smarts to get out of there!"[25.5]

He went on to show them a picture of his grandkids to underline the reason why we must act to save the species.

His laconic style is very refreshing, even though you have to listen carefully because you may miss the import in his throw-away statements and clipped low-energy delivery style:

"The Earth looks from the Moon, just like the Moon looks from the Earth [more or less] ... You can't really tell the difference at that distance. They all look very small at 240,000 miles out. But when people go to Mars, and they start looking back at two little dots out there after a few days, that ought to be an interesting trip."[25.1]

You'd better believe it! We have it on the authority of the most experienced NASA astronaut, one who went to the Moon twice, and one who left behind the most joyful of the lunar astronaut photos, a favorite at astronaut conventions, the "jump salute" photo. **Young** is saluting the flag and, in an exuberant moment, he demonstrates the 1/6-g environment as he leaps a couple of feet off the surface, while he is doing so. He does point out, however,[25.5] that photos can never give us the real experience:

"You know what it's like? It's just like taking a picture of a sunset on Earth—it never comes out half as good as when you saw it!"

As we have seen, very few of the Moon travelers have remained active enough to turn up at space development conferences now that they are in their eighties. But there is still one place where you can go to see a constellation of space explorers: Tucson, Arizona. There, amidst the red rocks and saguaro cactus, several of the Apollo veterans will assemble in a grand reunion, at Kim and Sally Poor's annual "Spacefest" astronaut autograph show and convention (Figure 6.34). What a strange kind of event this is. Certainly, there are sessions where scientists and authors give presentations about their space-related work. But also there is what you see demonstrated in Figure 6.35, where the astronauts (and others such as ground control veterans) from the Apollo era, sit at tables to sign their autographs for a range of fees which reflect their market values. There is something which is at the same time both distressing and wonderful about such a scene. On the one hand, it provides a venue when those who remain alive, and who have chosen to do so, can come and meet their old

Figure 6.34. Gathering of the Moon travelers. Here we see 11 of the 24 assembled in Tucson, Arizona in June, 2011. Amongst the group (see if you can find them!) are **Scott**, **Cernan**, **Duke**, **Haise**, **Stafford**, **Worden**, **Gordon**, **Lovell**, **Aldrin**, **Mitchell**, and **Bean**. We shall not see so many of them together again. Others in the image are astronauts that did not make it to the Moon.
Credit: Novaspace

Figure 6.35. Astronaut autograph signings still draw crowds. Moon travelers command premium prices.
Credit: MJMackowski

buddies again, nearly 50 years since those extraordinary events sent them off to visit our celestial neighbor. At the most recent such event, in June, 2017, there were now only 6 of the 24 who had made the Moon journey. Back in 2011, as we see in Figure 6.34, there were 11 of the 24 present. And, of course, that is precisely what presents the distressing part of the spectacle. The fact that very soon there will be nobody alive who made the trip, and who can therefore tell the story in person. It is a bit like watching the decline of the last remaining members of some magnificent species of snow leopard. As with all celebrities, they no doubt gradually get used to the interaction with the public. And are willing to satisfy the need for autographs. It is certainly a wonderful opportunity for any youngsters to be able to meet up with this rare breed, these "Living Legends," and maybe to be thereby motivated to go do their own thing afterwards, hopefully involving their own later space achievements. We have heard from several of the 24 Moon travelers who strongly support further space exploration, and heard their reasons why. This should be no surprise to us. However, several of them now realize that a new way will be needed, in order to make the journeys fundable, and we have engaged with them on possible ways to make the necessary progress—using re-usable technology, the **Aldrin** Cycler, and space tourism as the driver. In some cases, we also talked with

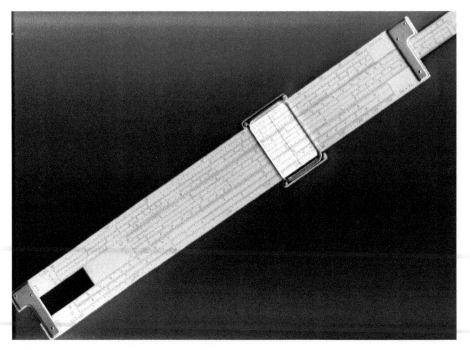

Figure 6.36. Memorabilia of the Golden Age of spaceflight. This is one of **Buzz Aldrin**'s full-size slide rules. See Figure 3.27 for the shorter version that he used in space.
Credit: Author's collection

them, and you heard their responses, about using prizes as a funding mechanism, with the Google Lunar XPRIZE as the example. Appendix D and Appendix E provide the framework for possibly moving forward.

There is an entire industry of "space memorabilia," which has grown around the mementos of the early space program (Figure 6.36), that you might term the relics of the space age. Autographs are of course a major part of the industry, with "Certificates of Authenticity" to support the provenance of objects and documents that have been signed by the Moon travelers. There are other astronauts at the show, too, but the Apollo veterans—those who went to the Moon—command premium prices. Even amongst the space memorabilia items there are graduations in value. If an item is "flown," then it commands a premium (provided, of course, that fact can be proven with documentation). Then, if it has been "flown to the Moon," it is worth significantly more. And the holy grail for collectors is "flown to the surface" material.

Anyone can be a collector of space memorabilia. Check out Robert Pearlman's collectspace.com website for all the news and possibilities. The cheapest items are probably postage stamps, or postal envelope covers, or newspapers, of the era. At the high end, space memorabilia auctions sometimes carry entire Soviet spacecraft, or "flown" spacesuits. And of course, there is nothing quite so rare, and pricy, as the proverbial Moon dust!

We move now to some analyses of the achievements of Apollo with 20/20 hindsight, and a half-century of data.   .

CHAPTER 7

Post-Flight Analysis

Half a century later, what can we learn from an analysis of the period of the 1960s, the whole Apollo-to- the-Moon program, and the lives of the 24 guys who made the Moon journey? We must remember that the entire process was a learning experience. Everything had to be made up as the program advanced. Each of the spacecraft was constructed as one of a series of ever-improving prototypes. All of the crewmen were test pilots plying their trade. And the ground control system, and all the training, had also to be established from scratch, with no prior examples to follow. This being the case, it is in retrospect astonishing that the program was so successful. **John Young** has said of it:

"It was really lucky that six of the missions did land on the Moon. We beat the odds." [25.5]

The two major accidents (Apollo 1 and Apollo 13) provided learning experiences for everyone involved.

Statistically, 24 men went to the Moon; 12 men walked on its surface; 12 orbited the Moon without landing; 6 orbited the moon solo; and 3 men conducted Lunar proximity EVAs. And nobody died during any of the spaceflights (Apollo 1 was a ground test failure). So, 24 out of the 24 men who went to the Moon returned safely to Earth—a 100% success rate. Sure, there were issues, and room for improvement, particularly with regard to the spacesuit design. There were some blind alleys followed during the test flights—in retrospect perhaps there was no need at that time to be conducting those dangerous Gemini gravity gradient tether experiments. And we were lucky to avoid fatalities attempting to fly the free-flying astronaut maneuvering units (AMUs), which again in retrospect did not have much to do with the conduct of the Moon landings, although of course, had they been successful, they could have provided an alternative to the Lunar Rover as a way of getting around on the lunar surface. There was maybe some poor feedback between the missions due to the compressed schedule, meaning that the same lessons had to be learned more than once. But, in general, the overall design, management, and system reliability throughout the Mercury/Gemini/Apollo era was remarkably good. And we must remember to include the humans in the loop when we think of the overall system. The 24 guys, without exception, delivered the goods. So, for the record, we should take a look at how they were recruited, what were the

Table 7.1. Astronaut recruitment statistics for the 24 Moon travelers.

Moon travelers		Astronaut group		Probabilities		
#	Names	#	Original qualified pool	Selected	Odds of getting into group	Percent of each group getting to Moon	Combined probability
1	Shepard	1	110	7	6%	14	0.8%
6	Armstrong, Borman, Lovell, Conrad, Stafford, Young	2	253	9	3.5%	66	2.3%
7	Bean, Collins, Gordon, Scott, Aldrin, Anders, Cernan	3	271	14	5.1%	50	2.5%
1	Schmitt	4	1351	6	0.4%	16	<0.1%
9	Haise, Swigert, Evans, Roosa, Mitchell, Irwin, Worden, Duke, Mattingly	5	351	19	5.4%	47	2.5%

Note: source of original pool numbers from [1.1]; for group recruitment criteria *see* Appendix B.

specifications, and how efficient was the process. All data as at July 2017, 48 years after the first Moon landing.

We have produced a table which summarizes this data on the recruitment process (Table 7.1). And the summarized specification for each wave of astronaut selection is provided in Appendix B, showing the changes introduced during the 7 years between the first and fifth group selections. Clearly, with the perspective looking back from 2017, we note a distinct lack of diversity in the selected candidates, reflecting the social framework of the US in the 1960s, and the starting point of military test pilots as the initial pool from which to select candidates. That having been said, what more can we learn from the post-flight analysis of the statistics?

First of all, we can see that, once selected as an astronaut in that era in groups 2 thru 5, there was a pretty good probability (around 50/50) that you would get to the Moon. So, in that sense Slayton did a good job in using his raw material and creating teams out of the astronauts, and training them. Group 4 was a little different—and as we have seen "the scientists" were not given the same priority as the traditional jet pilots in getting Moon flight shots. There was not much difference from group to group, so the slight changes to the selection criteria did not unduly influence the outcome. The other useful data point is that, given the size of the original candidate pools, you had about a 5% chance of becoming one of the selected astronauts in each tranche. So, these guys were in the top 5% of their peer group. Again, in the case of the scientists, it was a much harder hurdle to become selected. If we combine the two probabilities once you make the potential pool of candidates with the required qualifications, that is, first of all being selected as an astronaut, and then secondly to be chosen as prime crew on a Moon mission, we find the odds were only around 2.5% that you would get to the Moon. So, no surprises there. We knew they were pretty special, and they knew it, too!

We have seen how the selection process worked. What can we learn in retrospect regarding the career statistics of the 24? We provide summary information in Figure 7.1.

We show three different perspectives on the astronaut careers of the Moon travelers. We note that each of the 24 recruited into NASA stayed at least 6 years, half of them left after 10 years, after the Moon landings were over, and half stayed on. All except **Young** had left before their 20th year at the agency. They were aged about 38 when they went to the Moon, except for "old man" **Shepard**, the only one of the first astronaut group to make it to the Moon. The need for an accelerated program while trying to meet Kennedy's challenge clearly resulted in some inefficiencies, with nearly half of the Moon traveling astronauts leaving the agency after only one flight, something which certainly made it an expensive flight, when the costs of training and the salaries for up to 20 years at NASA are considered. Even with 4 flights in a 20-year period, it hardly makes for an efficient use of the investment in astronaut training. Again, **Young** is something of an anomaly with his six flights, but he spent twice as long as anyone else at NASA, mostly in non-flying senior desk jobs. In hindsight, we can see that this was a very focused period, those involved in making it happen did what they had to do, and succeeded. But some inefficiencies are inevitable with that kind of time-constrained target with unconstrained financial resources (around 5% of GDP). The message for the future is clear. For a sustainable human spaceflight program, with constrained financial resources, there needs to be a plan for ongoing space exploration, which allows more use of the selected astronauts on a regular basis, which does not mean only an average of flying once every 5 years!

The 24 Moon travelers were selected, as we have seen, through a process which emphasized excellent physical conditioning. And yet, we have noted in the narrative that several of the astronauts were dogged by various medical issues during their time at NASA. For instance, Slayton, from the original Mercury Seven group, was subsequently found to have a heart murmur. **Shepard** had his balancing problem—Menière's disease. **Lovell** had some kind of liver problem. **Collins** had a bone spur in his neck vertebrae. **Irwin** also had heart problems. So, one supposes that, even though the tests were extensive, the medical checks during the selection process missed some indicators. One suspects that by today's standards, such tests would result in a more thorough vetting. Having said that, let me hastily make clear that I for one am glad that

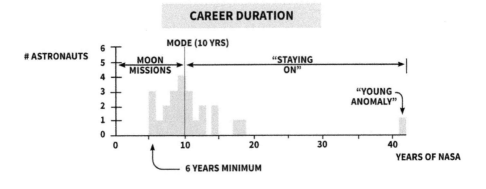

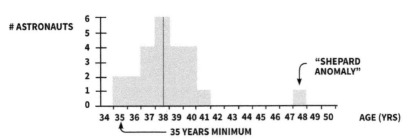

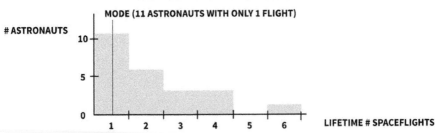

Figure 7.1. Astronaut career statistics for the 24 Moon travelers.

all of the above did succeed in being recruited—they all served admirably during their missions. The size of the data pool is too small for any meaningful statistics to be derived, but Figure 7.2 has been provided to summarize the aggregate data on mortality, for what it is worth.

At the time of writing (July 1, 2017), 9 of the 24 had died (with the individual data recorded in Appendix A for each Moon traveler). In general, the group has turned out to be long-lived, as one would expect and hope given

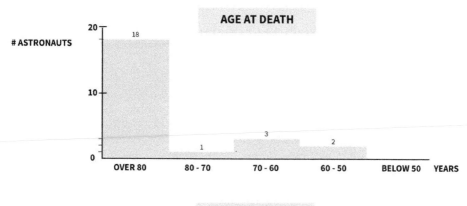

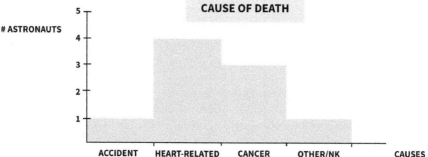

Figure 7.2. Astronaut mortality statistics for the 24 Moon travelers. Note: based on 9 deaths by mid-2017.

their health at the outset of their careers, with 18 of the 24 living into their eighties (and maybe eventually some into their nineties). Fifteen remain alive at this time, all currently at least 83 in age. We have too little data to conclude anything from the causes of death, with heart problems and cancer being common causes in the population at large. And we see no correlation between the causes of death and the number or duration of flights. So, that is the raw hard data. What can we tell in aggregate from more "soft" data? What do we learn from their collective thoughts and insights? There is nothing we can tabulate, but some impressions are worth noting.

Like does going to the Moon change a person? At least **Irwin**, **Duke**, and **Mitchell** found the experience dramatically life-changing. Others, like **Bean**, **Gordon**, and **Roosa**, also testified to the significant impact of their Moon trips. So, right there, we have a quarter of the travelers reporting the significance. And, if truth be known, I suspect that we would have a 100% response were it not for the need of jet-jocks to "remain cool" about such things.

Going to the Moon is a tough act to follow. What did the 24 Moon travelers do with the rest of their lives when they returned to Earth, still aged only between 35 and 41? One of the characteristics in common was that they all faced challenges in deciding what to do, and many of them changed about from one kind of job to another over time. However, we can state that, for at least part of their remaining working lives, 4 out of 24 tried politics, 7 out of 24 tried consulting, only 3 out of 24 continued flying, 4 out of 24 tried the beer business, 14 out of 24 became authors, and 2 out of 24 became religious leaders. The majority (14 out of 24) made their living post-Apollo as corporate executives and board members. I can add, from personal observation, that almost without exception, and despite this not being part of their initial vetting process, they all became good presenters, thoughtful interviewees, maintained a competitiveness and sense of humor, and have been wonderful ambassadors for America and the space program.

And, what do they have to tell us, collectively, about the future of the space program? They all have become accepting that there will be no further US human Moon landings in their remaining lifetimes, or at least no human footprints on Mars. We address going back to the Moon in Appendix D, and going further in Appendix E. Maybe a half of the Moon travelers have clearly stated the need for more exploration and even settlement (**Armstrong, Aldrin, Scott, Duke, Collins, Gordon, Haise, Mattingly, Young**, and **Schmitt**). After initial skepticism, maybe the majority of the 24 now accept, and welcome, the benefits that space tourism, and commercial space in general, will bring. Some (**Conrad, Mattingly**, and **Aldrin**) were actively involved in trying to bring about changes from the government-funded NASA approach to a future period of commercial space exploration.

Finally, when all is said and done, the future belongs to the children, and in this case—50 years on—to the grandchildren, and to all the succeeding generations. How well did they do in providing a future generation, leading to descendants of the Moon travelers? It turns out they score very highly in the "Mayflower Effect" league. Good job, guys! The children of the 24 are today aged between 33 and 70 years, and there are a multitude of grandchildren. This is collectively how the 24 Moon travelers "scored" in terms of numbers of children: the average number was almost exactly 3 children per Moon traveler, with only 2 of the astronauts being childless, and 3 of them each having 6 children. So the legacy is assured!

CHAPTER 8

Conclusions

Twenty-four guys went to the Moon around the end of the 1960s. The lack of any subsequent attempts to repeat their journeys in the next half century, by astronauts from any country, only serves to underline how difficult and risky was that early endeavor. And of course how expensive. Clearly, Apollo was an anomaly, and Moon journeys will not be repeatable in that same way. We need to find new ways if we are going to venture back to the Moon, or continue farther to other celestial objects like Mars. And these new ways will need to embody a commercial incentive. Towards the end of their lives, many of the Apollo Moon travelers began to recognize that fact. Although for decades they had hoped to see new explorers follow in their footsteps following a new "Kennedy narrative." They all experienced a decompression after the excitement and high pace of activities of Apollo came to an end. Some coped better than others, but I suspect every one of them was surprised when there was no follow-up mission, and indeed when the American public itself lost interest. So, what was all the rush, and what, if any, is the legacy of all the risky missions undertaken by the 24 guys who flew to the Moon? We have explained the compressed timescales which were due to JFK's decision to use space travel to the Moon as a yardstick to measure the pros and cons of a Western free-economy versus a Soviet communist command economy. We know the outcome. Apollo helped the West to win the Cold War.

What was the more lasting legacy, though?

We are familiar with the spin-off theory, and know that it is undeniably true that Apollo kick-started a great many industries, and even new directions in science and engineering, creating new businesses such as micro-electronics, that would end up transforming our daily lives. But there is more.

One major legacy was the birth of the "protect the Earth" environmental movement. Arguably the photos they brought back, and the travelers' own *ad hoc* comments as recorded in this book, were the main contribution to changing the sensitivities of humanity to the recognition of the special place that is our celestial home. And, in addition, a renewed focus is emerging on eventually having a back-up plan for life elsewhere, given the fragility of our home planet in the grand scheme of things.

Furthermore, some of the astronauts also made their own distinct contributions to ongoing space exploration and science after their return to Earth, so that we might continue the exploration that they started. **Aldrin** has given us his Cycler idea, which will help pave the way to repeatable interplanetary travel. **Schmitt** has helped us focus on the potential for mining the Moon for its Helium-3, which could provide vast energy potential for us here on Earth. **Mitchell** devoted his life to creating the framework for a new science of Noetics—which though still in its infancy might bring new insights into how our brains work. We have also offered some other pieces of the puzzle, and in the book provided the responses of some of the Moon travelers to such enterprises as space tourism, the commercial use of the Moon (Google Lunar XPRIZE), and to the proposed Gateway Earth architecture for repeatable relatively low-cost interplanetary travel.

One significant irreversible and lasting legacy of the Apollo missions is the completion of the sentence: "If they can go to the Moon, then ... (add your own grand challenge)." Nowadays, that notion is universal, and the Internet has certainly expanded the possibilities enormously. Maybe the greatest legacy of Apollo has been the creation and motivation of a whole generation of space entrepreneurs (Musk, Bezos, Branson, Bigelow, etc.) who are defining the new way forward (see Appendices D and E), fueled by the amazing experiences they recall, and the knowledge that such extraordinary things are possible, from their childhood when Man went to the Moon. Jacob Bronowski ended his great 1973 work *The Ascent of Man*, with this distilled essence of everything he had learned in his distinguished life:

"Every man, every civilization, has gone forward because of its engagement with what it has set itself to do. The personal commitment of a man to his skill, the intellectual commitment and the emotional commitment working together as one, has made the Ascent of Man."

So, maybe old Gus Grissom had it right all along, when asked to address the workers who were building his rocket:

"Just do good work."

The main legacy of these 24 chosen men is then, just as it is for all of us, in the good work they achieved through their commitment to their profession. And in the people who have been affected by their presence and contributions—both during that amazing Apollo era, and subsequently. They risked their lives, and as a consequence gave mankind a future of unlimited possibilities. We owe it to them to continue their task.

SECTION III

Appendices

"It's important that we attempt to extend life beyond Earth now. It is the first time in the four billion-year history of Earth that it's been possible, and that window could be open for a long time—hopefully it is—or it could be open for a short time. We should err on the side of caution and do something now."

Elon Musk
Founder and CEO, SpaceX

Moon Traveler Basic Data

ASTRONAUT: ALDRIN, EDWIN EUGENE ("BUZZ")

(credit: NASA)

Summary lunar experience

Apollo 11, Lunar Module Pilot, third mission to the Moon, first landing mission

Personal data

Date of birth	January 30, 1930
Place of birth	Montclair, New Jersey
Married	Joan (1954), Beverly (1975), Lois (1987)
Children	Michael (1955), Janice (1957), Andrew (1958)
Death—date/cause/age	N/A

Military history

Branch	USAF
Rank	Colonel
Flying experience	F-86s (Korea), F-100s (Germany)

Academic/Engineering qualifications

Degree	B.Sc.	Sc.D.
Subject	Military Science	Astronautics
Place	West Point	MIT

NASA experience

Recruitment date/Group	October, 1963/Group 3, "The Fourteen"
Space missions	Gemini 12, Apollo 11
Retirement date	July 1971

(credit: NASM)

ASTRONAUT: ANDERS, WILLIAM ALISON

(credit: NASA)

Summary lunar experience

Apollo 8, Lunar Module Pilot, first mission to the Moon

Personal data

Date of birth	Oct 17, 1933
Place of birth	Hong Kong (British Mandate)
Married	Valerie (1955)
Children	Alan (1957), Glen (1958), Gayle (1960), Gregory (1962), Eric (1964), Diana (1972)
Death—date/cause/age	N/A

Military history

Branch	USAF
Rank	Major
Flying experience	F-89s (Iceland)

Academic/Engineering qualifications

Degree	B.Sc.	M.Sc.
Subject	Military Science	Nuclear Engineering
Place	Annapolis	USAF Inst of Technology

NASA experience

Recruitment date/Group	October, 1963/Group 3, "The Fourteen"
Space missions	Apollo 8
Retirement date	1973

(credit: Heritage Flight Museum)

ASTRONAUT: ARMSTRONG, NEIL ALDEN

(credit: NASA)

Summary lunar experience

Apollo 11, Commander, third mission to the Moon, first landing mission

Personal data

Date of birth   Aug 5, 1930
Place of birth   Wapakoneta, Ohio
Married   Janet (1956), Carol (1994)
Children   Eric (1957), Karen (1959), Mark (1963)
Death—date/cause/age   Aug 25, 2012/heart bypass surgery complications/82

Military history

Branch   Civilian (former US Navy)
Rank   N/A
Flying experience   Naval aviator, Panther (Korea), F-104, X-1B, X-15
                       (Edwards)

Academic/Engineering qualifications

Degree   B.Sc.   M.Sc.
Subject   Aeronautical Engineering Aerospace Engineering
Place   Purdue   USC (LA)

NASA experience

Recruitment date/Group   June, 1962/Group 2, "The New Nine"
Space missions   Gemini 8, Apollo 11
Retirement date   1971

(credit: NASM)

ASTRONAUT: BEAN, ALAN LAVERN

(credit: NASA)

Summary lunar experience

Apollo 12, Lunar Module Pilot, fourth mission to the Moon, second landing mission

Personal data

Date of birth	March 15, 1932
Place of birth	Wheeler, Texas
Married	Sue (1955), Leslie (1982)
Children	Clay (1955), Amy (1963)
Death—date/cause/age	N/A

Military history

Branch	US Navy
Rank	Captain
Flying experience	Naval aviator, F-4H Phantoms (Jacksonville, Pax River)

Academic/Engineering qualifications

Degree	B.Sc.
Subject	Aeronautical Engineering
Place	Texas

NASA experience

Recruitment date/Group	October, 1963/Group 3, "The Fourteen"
Space missions	Apollo 12, Skylab 2
Retirement date	June 1981

(credit: Russo/NASM)

ASTRONAUT: BORMAN, FRANK

(credit: NASA)

Summary lunar experience

Apollo 8, Commander, first mission to the Moon

Personal data

Date of birth	March 14, 1928
Place of birth	Gary, Indiana
Married	Susan (1950)
Children	Frederick (1951), Edwin (1953)
Death—date/cause/age	N/A

Military history

Branch	USAF
Rank	Colonel
Flying experience	F-86 Sabre, instructor at Edwards

Academic/Engineering qualifications

Degree	B.Sc.	M.Sc.
Subject	Military Science	Aeronautical Engineering
Place	West Point	Caltech

NASA experience

Recruitment date/Group	June, 1962/Group 2, "The New Nine"
Space missions	Gemini 7, Apollo 8
Retirement date	1970

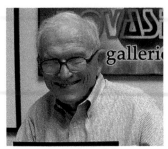

(credit: Novaspace Galleries)

ASTRONAUT: CERNAN, EUGENE ANDREW

(credit: NASA)

Summary lunar experience

Apollo 10, Lunar Module Pilot, second mission to the Moon Apollo 17, Commander, ninth mission to Moon, sixth landing mission

Personal data

Date of birth	March 14, 1934
Place of birth	Chicago, Illinois
Married	Barbara (1961), Jan (1987)
Children	Teresa (1963) and two stepdaughters
Death—date/cause/age	January 16, 2017/not announced/82

Military history

Branch	US Navy
Rank	Captain
Flying experience	Naval aviator, A-4 Skyhawk, F9F Panther

Academic/Engineering qualifications

Degree	B.Sc.	M.Sc.
Subject	Electrical Engineering	Aeronautical Engineering
Place	Purdue	US Navy Postgrad School

NASA experience

Recruitment date/Group	October, 1963/Group 3, "The Fourteen"
Space missions	Gemini 9, Apollo 10, Apollo 17
Retirement date	July, 1976

(credit: Author)

ASTRONAUT: COLLINS, MICHAEL

(credit: NASA)

Summary lunar experience

Apollo 11, Command Module Pilot, third mission to Moon, first landing mission

Personal data

Date of birth	October 31, 1930
Place of birth	Rome, Italy
Married	Pat (1957)
Children	Kathleen (1959), Ann (1961), Mike (1963)
Death—date/cause/age	N/A

Military history

Branch	USAF
Rank	Lieutenant Colonel
Flying experience	F-86s (Iceland), test pilot at Edwards

Academic/Engineering qualifications

Degree	B.Sc.
Subject	Military Science
Place	West Point

NASA experience

Recruitment date/Group	October, 1963/Group 3, "The Fourteen"
Space missions	Gemini 10, Apollo 11
Retirement date	January 1970

(credit: NASM)

ASTRONAUT: CONRAD, CHARLES ("PETE")

(credit: NASA)

Summary lunar experience

Apollo 12, Commander, fourth mission to the Moon, second landing mission

Personal data

Date of birth	June 2, 1930
Place of birth	Philadelphia, Pennsylvania
Married	Jane (1953), Nancy (1990)
Children	Peter (1954), Thomas (1957), Andrew (1959), Christopher (1960)
Death—date/cause/age	July 8, 1999/motorcycle accident/69

Military history

Branch	US Navy
Rank	Captain
Flying experience	Naval aviator, F-4H Phantoms, Pax River test pilot

Academic/Engineering qualifications

Degree	B.Sc.
Subject	Aeronautical Engineering
Place	Princeton

NASA experience

Recruitment date/Group	June, 1962/Group 2, "The New Nine"
Space missions	Gemini 5, Gemini 11, Apollo 12, Skylab 1
Retirement date	1973

(credit: Steve Pyke)

ASTRONAUT: DUKE, CHARLES MOSS

(credit: NASA)

Summary lunar experience

Apollo 16, Lunar Module Pilot, eighth mission to the Moon, fifth landing mission

Personal data

Date of birth	October 3, 1935
Place of birth	Charlotte, North Carolina
Married	Dorothy (1963)
Children	Charles (1965), Thomas (1967)
Death—date/cause/age	N/A

Military history

Branch	USAF
Rank	Lieutenant Colonel
Flying experience	F-86L (Germany), flight instructor at Edwards

Academic/Engineering qualifications

Degree	B.Sc.	M.Sc.
Subject	Naval Sciences	Aeronautics and Astronautics
Place	Annapolis	MIT

NASA experience

Recruitment date/Group	April, 1966/Group 5, "The Original Nineteen"
Space missions	Apollo 16
Retirement date	January 1976

(credit: Zimbio.com)

ASTRONAUT: EVANS, RONALD ELLWIN

(credit: NASA)

Summary lunar experience

Apollo 17, Command Module Pilot, ninth mission to the Moon, sixth landing mission

Personal data

Date of birth   November 10, 1933
Place of birth   St. Francis, Kansas
Married   Janet (1957)
Children   Jaime (1959), Jon (1961)
Death—date/cause/age   April 7, 1990/heart attack/56

Military history

Branch   US Navy
Rank   Captain
Flying experience   Naval aviator (Vietnam) F-8, flight instructor

Academic/Engineering qualifications

Degree   B.Sc.   M.Sc.
Subject   Electrical Engineering   Aeronautical Engineering
Place   Kansas   US Navy Postgrad School

NASA experience

Recruitment date/Group   April, 1966/Group 5, "The Original Nineteen"
Space missions   Apollo 17
Retirement date   March 1977

(credit: photodivision.gov.in)

ASTRONAUT: GORDON, RICHARD FRANCIS

(credit: NASA)

Summary lunar experience

Apollo 12, Command Module Pilot, fourth mission to the Moon, second landing mission

Personal data

Date of birth	October 5, 1929
Place of birth	Seattle, Washington
Married	Barbara (1953), Linda (1981)
Children	Carleen (1954), Richard (1955), Lawrence (1957), Thomas (1959), James (1960), Diane (1961)
Death—date/cause/age	N/A

Military history

Branch	US Navy
Rank	Captain
Flying experience	Naval aviator, F-4H Phantoms, Pax River Flight Safety Officer

Academic/Engineering qualifications

Degree	B.Sc.
Subject	Chemistry
Place	Seattle

NASA experience

Recruitment date/Group	October, 1963/Group 3, "The Fourteen"
Space missions	Gemini 11, Apollo 12
Retirement date	January 1972

(credit: Explorationatitsgreatest.blogspot.com)

ASTRONAUT: HAISE, FRED WALLACE

(credit: NASA)

Summary lunar experience

Apollo 13, Lunar Module Pilot, fifth mission to the Moon

Personal data

Date of birth	November 17, 1933
Place of birth	Biloxi, Mississippi
Married	Mary (1956), Patt (1979)
Children	Mary (1956), Frederick (1958), Stephen (1961), Thomas (1970)
Death—date/cause/age	N/A

Military history

Branch	Civilian (former Marine and USAF)
Rank	N/A
Flying experience	Edwards test pilot (multiple aircraft types)

Academic/Engineering qualifications

Degree	B.Sc.
Subject	Aerospace Engineering
Place	Oklahoma

NASA experience

Recruitment date/Group	April, 1966/Group 5, "The Original Nineteen"
Space missions	Apollo 13, Shuttle (approach and landing tests)
Retirement date	June 1979

(credit: Mark Usciak/Spaceflight Insider)

ASTRONAUT: IRWIN, JAMES BENSON

(credit: NASA)

Summary lunar experience

Apollo 15, Lunar Module Pilot, seventh mission to the Moon, fourth landing mission

Personal data

Date of birth	March 17, 1930
Place of birth	Pittsburgh, Pennsylvania
Married	Mary (1959)
Children	Joy (1959), Jill (1961), James (1963), Jan (1964), Joe (1969)
Death—date/cause/age	August 8, 1991/heart attack/61

Military history

Branch	USAF
Rank	Colonel
Flying experience	YF 12A, test pilot at Edwards

Academic/Engineering qualifications

Degree	B.Sc.	M.Sc.	M.Sc.
Subject	Naval Science	Aeronautical Engineering	Instrumentation Engineering
Place	Annapolis	Michigan	Michigan

NASA experience

Recruitment date/Group	April, 1966/Group 5, "The Original Nineteen"
Space missions	Apollo 15
Retirement date	July 1972

(credit: www.stripes.com)

ASTRONAUT: LOVELL, JAMES ARTHUR

(credit: NASA)

Summary lunar experience

Apollo 8, Command Module Pilot, first mission to the Moon, Apollo 13, Commander, fifth mission to the Moon

Personal data

Date of birth   March 25, 1928
Place of birth   Cleveland, Ohio
Married   Marilyn (1952)
Children   Barbara (1953), James (1955), Susan (1958), Jeffrey (1966)
Death—date/cause/age   N/A

Military history

Branch   US Navy
Rank   Captain
Flying experience   Naval aviator, F-4H Phantom, Pax River test pilot

Academic/Engineering qualifications

Degree   B.Sc.
Subject   Military Science
Place   Annapolis

NASA experience

Recruitment date/Group   June, 1962/Group 2, "The New Nine"
Space missions   Gemini 7, Gemini 12, Apollo 8, Apollo 13
Retirement date   March 1973

(credit: Blog.Sciencemuseum.org.uk)

ASTRONAUT: MATTINGLY, THOMAS KENNETH

(credit: NASA)

Summary lunar experience

Apollo 16, Command Module Pilot, eighth mission to the Moon, fifth landing mission

Personal data

Date of birth	March 17, 1936
Place of birth	Chicago, Illinois
Married	Elisabeth (1970)
Children	Thomas (1972)
Death—date/cause/age	N/A

Military history

Branch	US Navy
Rank	Rear Admiral
Flying experience	Naval aviator (A-1H Skyraider), Edwards (USAF) test pilot

Academic/Engineering qualifications

Degree	B.Sc.
Subject	Aeronautical Engineering
Place	Auburn

NASA experience

Recruitment date/Group	April, 1966/Group 5, "The Original Nineteen"
Space missions	Apollo 16, Shuttle STS-4, Shuttle STS-51C
Retirement date	1985

(credit: www.Spacelectures.com)

ASTRONAUT: MITCHELL, EDGAR DEAN

(credit: NASA)

Summary lunar experience

Apollo 14, Lunar Module Pilot, sixth mission to the Moon, third landing mission

Personal data

Date of birth	September 17, 1930
Place of birth	Hereford, Texas
Married	Louise (1951), Anita (1973), Sheilah (1989)
Children	Karlyn (1953), Elisabeth (1959), Kimberly (1961), Paul (1963), Mary (1964), Adam (1984)
Death—date/age	February 4, 2016/not reported/ 85

Military history

Branch	US Navy
Rank	Captain
Flying experience	Naval aviator, A-3D Skywarrior (Japan), instructor at Edwards

Academic/Engineering qualifications

Degree	B.Sc.	B.Sc.	Sc.D.
Subject	Industrial Management	Aeronautics	Aeronautics and Astronautics
Place	Carnegie Mellon	Annapolis	MIT

NASA experience

Recruitment date/Group	April, 1966 / Group 5, "The Original Nineteen"
Space missions	Apollo 14
Retirement date	1972

(credit: Author)

ASTRONAUT: ROOSA, STUART ALLEN

(credit: NASA)

Summary lunar experience

Apollo 14, Command Module Pilot, sixth mission to the Moon, third landing mission

Personal data

Date of birth	August 16, 1933
Place of birth	Durango, Colorado
Married	Joan (1957)
Children	Christopher (1959), John (1961), Stuart (1962), Rosemary (1963)
Death—date/cause/age	December 12, 1994/pancreatitis/61

Military history

Branch	USAF
Rank	Colonel
Flying experience	F-84F, F-100, F-101 (Japan), test pilot at Edwards

Academic/Engineering qualifications

Degree	B.Sc.
Subject	Aeronautical Engineering
Place	Boulder

NASA experience

Recruitment date/Group	April, 1966/Group 5, "The Original Nineteen"
Space missions	Apollo 14
Retirement date	1976

(credit: Members.chello.at)

ASTRONAUT: SCHMITT, HARRISON HAGAN ("JACK")

(credit: NASA)

Summary lunar experience

Apollo 17, Lunar Module Pilot, ninth mission to the Moon, sixth landing mission

Personal data

Date of birth	July 3, 1935
Place of birth	Santa Rita, New Mexico
Married	Theresa (1985)
Children	None
Death—date/cause/age	N/A

Military history

Branch	Civilian
Rank	N/A
Flying experience	1-year of flight school (Cessna 172, T-38, helicopters)

Academic/Engineering qualifications

Degree	B.Sc.	Ph.D.
Subject	Geology	Geology
Place	Caltech	Harvard

NASA experience

Recruitment date/Group	June, 1965/Group 4, "The Scientists"
Space missions	Apollo 17
Retirement date	1976

(credit: Author)

ASTRONAUT: SCOTT, DAVID RANDOLPH

(credit: NASA)

Summary lunar experience

Apollo 15, Commander, seventh mission to the Moon, fourth landing mission

Personal data

Date of birth	June 6, 1932
Place of birth	San Antonia, Texas
Married	Lurton (1959), Margaret (2005)
Children	Tracy (1961), Douglas (1963)
Death—date/cause/age	N/A

Military history

Branch	USAF
Rank	Colonel
Flying experience	F-86 and F-100 (Netherlands), test pilot at Edwards

Academic/Engineering qualifications

Degree	B.Sc.	M.Sc .
Subject	Military Science	Aeronautics and Astronautics
Place	West Point	MIT

NASA experience

Recruitment date/Group	October, 1963/Group 3, "The Fourteen"
Space missions	Gemini 8, Apollo 9, Apollo 15
Retirement date	1978

(credit: Author)

ASTRONAUT: SHEPARD, ALAN BARTLETT

(credit: NASA)

Summary lunar experience

Apollo 14, Commander, sixth mission to the Moon, third landing mission

Personal data

Date of birth	November 18, 1923
Place of birth	East Derry, New Hampshire
Married	Louise
Children	Laura (1947), Julie (1951), (and niece Alice)
Death—date/cause/age	July 21, 1998/leukemia/74

Military history

Branch	US Navy
Rank	Rear Admiral
Flying experience	Naval aviator, F-3H, F-5D (Mediterranean), instructor at Pax River

Academic/Engineering qualifications

Degree	B.Sc.
Subject	Naval Science
Place	Annapolis

NASA experience

Recruitment date/Group	April, 1959/Group 1, "The Mercury Seven"
Space missions	Mercury "Freedom Seven," Apollo 14
Retirement date	August 1974

(credit: Author)

ASTRONAUT: STAFFORD, THOMAS PATTEN

(credit: NASA)

Summary lunar experience

Apollo 10, Commander, second mission to the Moon

Personal data

Date of birth	September 17, 1930
Place of birth	Weatherford, Oklahoma
Married	Faye (1953), Linda (1988)
Children	Dionne (1954), Karin (1957), Michael (adopted), Stanislav (adopted)
Death—date/cause/age	N/A

Military history

Branch	USAF
Rank	Lieutenant General
Flying experience	F-86D (Germany), instructor at Pax River and Edwards

Academic/Engineering qualifications

Degree	B.Sc.
Subject	Military Science
Place	Annapolis

NASA experience

Recruitment date/Group	June, 1962/Group 2, "The New Nine"
Space missions	Gemini 6, Gemini 9, Apollo 10, ASTP
Retirement date	August 1975

(credit: Stafford Museum)

ASTRONAUT: SWIGERT, JOHN LEONARD ("JACK")

(credit: NASA)

Summary lunar experience

Apollo 13, Command Module Pilot, fifth mission to the Moon

Personal data

Date of birth	August 30, 1931
Place of birth	Denver, Colorado
Married	No
Children	No
Death—date/cause/age	December 27, 1982/bone cancer/51

Military history

Branch	USAF
Rank	Captain
Flying experience	F 100-A Supersabre (Japan, Korea), commercial test pilot

Academic/Engineering qualifications

Degree	B.Sc.	M.Sc.	M.B.A.
Subject	Mechanical Engineering	Aeronautical Science	Business
Place	Colorado	New York	Hartford

NASA experience

Recruitment date/Group	April, 1966/Group 5, "The Original Nineteen"
Space missions	Apollo 13
Retirement date	August, 1977

(credit: Duane Howell)

ASTRONAUT: WORDEN, ALFRED MERRILL

(credit: NASA)

Summary lunar experience

Apollo 15, Command Module Pilot, seventh mission to the Moon, fourth landing mission

Personal data

Date of birth	February 7, 1932
Place of birth	Jackson, Michigan
Married	Pamela (1955), Sandra (1974), Jill (1982)
Children	Tamara (1956), Merrill (1958), Alison (1960),
Death—date/cause/age	N/A

Military history

Branch	USAF
Rank	Lieutenant Colonel
Flying experience	F-86, F-102, F-106 (Andrews AFB), Meteors, Hunters (test pilot at Farnborough), F-104 (instructor at Edwards)

Academic/Engineering qualifications

Degree	B.Sc.	M.Sc.	M.Sc.
Subject	Military Science	Aero and Astro Engineering	Instrumentation Engineering
Place	West Point	Michigan	Michigan

NASA experience

Recruitment date/Group	April, 1966/Group 5, "The Original Nineteen"
Space missions	Apollo 15
Retirement date	1975

(credit: Collectspace.com)

ASTRONAUT: YOUNG, JOHN WATTS

(credit: NASA)

Summary lunar experience

Apollo 10, Command Module Pilot, second mission to the Moon Apollo 16, Commander, eighth mission to the Moon, fifth landing mission

Personal data

Date of birth	September 24, 1930
Place of birth	San Francisco, California
Married	Barbara (1955), Susy (1972)
Children	Sandy (1957), John (1959)
Death—date/cause/age	N/A

Military history

Branch	US Navy
Rank	Captain
Flying experience	Naval aviator (Korea), Cougars, Crusaders, Phantom, Pax River test pilot

Academic/Engineering qualifications

Degree	B.Sc.
Subject	Aeronautical Engineering
Place	Georgia Tech

NASA experience

Recruitment date/Group	June, 1962/Group 2, "The New Nine"
Space missions	Gemini 3, Gemini 10, Apollo 10, Apollo 16, Shuttle STS-1, Shuttle STS-9
Retirement date	2004

(credit: Novaspace Galleries)

Astronaut Entry Groups

Moon travelers indicated **emboldened** (photos credit: NASA)
(Main sources: 1.1, 1.15)

GROUP 1, APRIL 1959, "THE MERCURY SEVEN" (1 OF 7 BECAME A MOON TRAVELER)

(*Left to right*) Scott Carpenter, Gordon Cooper, John Glenn, Gus Grissom, Wally Schirra, **Alan Shepard**, Deke Slayton.

Criteria:
- US citizen.
- Male.
- Aged 25–40 years.
- Less than 5 ft 11 in. tall.
- In excellent physical condition.
- Holds a Bachelor's Degree in Mathematics, Science, or Engineering.
- A jet pilot with at least 1,500 hours flying time.
- A test pilot and/or combat mission experience.

GROUP 2, JUNE 1962, "THE NEW NINE" (6 OF 9 BECAME MOON TRAVELERS)

(*Back row*) See, McDivitt, **Lovell**, White, **Stafford**. (*Front row*) **Conrad**, **Borman**, **Armstrong**, **Young**.

Criteria:

- US citizen.
- Male.
- Aged 25–35 years.
- Less than 6 ft.
- In excellent physical condition.
- Holds a Bachelor's degree in Physical or Biological Sciences, or Engineering,
- A test pilot (military, aircraft industry, or NASA).

(Notes: height increased due to larger Gemini; age dropped to allow for all upcoming Apollo flights.)

GROUP 3, OCTOBER 1963, "THE FOURTEEN" (7 OF 14 BECAME MOON TRAVELERS)

(*Back row*) **Collins**, Cunningham, Eisele, Freeman, **Gordon**, Schweickart, **Scott**, and Williams. (*Front row*) **Aldrin**, **Anders**, Bassett, **Bean**, **Cernan**, and Chaffee.

Criteria:
- US citizen.
- Aged 25–34 years.
- Less than 6 ft.
- In excellent physical condition.
- Holds a Bachelor's Degree in Physical or Biological Sciences, or Engineering.
- Has 1,000 hours of flight time with military, aircraft industry, or NASA.

(Note: no test flying requirement—Slayton decided that they had used up that pool, and opened up to non-test pilots.)

GROUP 4, JUNE 1965, "THE SCIENTISTS" (1 OF 5 BECAME A MOON TRAVELER)

(*Back row*) Garriott and Gibson. (*Front row*) Michel, **Schmitt**, and Kerwin.

Criteria:
- US citizen.
- Aged at least 34.
- Less than 6 ft.
- Holds a Ph.D. in natural sciences, medicine, or engineering.

(Note: no flying experience necessary.)

GROUP 5, APRIL 1966, "THE ORIGINAL NINETEEN" (9 OF 19 BECAME MOON TRAVELERS)

(*Back row*) **Swigert**, Pogue, **Evans**, Weitz, **Irwin**, Carr, **Roosa**, **Worden**, **Mattingly**, and Lousma. (*Front row*) Givens, **Mitchell**, **Duke**, Lind, **Haise**, Engle, Brand, Bull, and McCandless.

Criteria:
- US citizen.
- Aged 25–36 years.
- Less than 6 ft.
- In excellent physical condition.
- Holds a Bachelor's Degree in Physical or Biological Sciences, or Engineering.
- Has 1,000 hours of flight time with military, aircraft industry, or NASA.

Key Mission Summaries

Part 1

Spacecraft	Designation	Launch date	Crew	Remarks
Mercury	MR-3	May 5, 1961	**Shepard**	1st US spaceflight—suborbital
Mercury	MR-4	July 21, 1961	Grissom	Completed suborbital testing
Mercury	MA-6	February 20, 1962	Glenn	1st American in orbit—3 orbits
Mercury	MA-7	May 24, 1962	Carpenter	3 orbits—scientific experiments
Mercury	MA-8	October 3, 1962	Schirra	6 orbits duration
Mercury	MA-9	May 15, 1963	Cooper	22 orbits duration
Gemini	GT-3	March 23, 1965	Grissom, **Young**	1st Gemini test
Gemini	GT-4	June 3, 1965	McDivitt, White	1st US EVA
Gemini	GT-5	August 21, 1965	Cooper, **Conrad**	8 days duration
Gemini	GT-6A	December 15, 1965	Schirra, **Stafford**	1st rendezvous
Gemini	GT-7	December 4, 1965	**Borman**, Lovell	14 days duration
Gemini	GT-8	March 16, 1966	**Armstrong**, Scott	1st docking in orbit
Gemini	GT-9A	June 3, 1966	**Stafford**, Cernan	Multiple rendezvous, EVA
Gemini	GT-10	July 18, 1966	**Young**, Collins	Rendezvous, dock, fired Agena motor
Gemini	GT-11	September 12, 1966	**Conrad**, Gordon	Rendezvous, dock, EVAs, fired Agena, Tether
Gemini	GT-12	November 11, 1966	**Lovell**, Aldrin	Rendezvous, dock, EVAs, Tether

Part 2

Spacecraft	Designation	Launch date	Crew	Remarks
Apollo	7	October 11, 1968	Schirra, Eisele, Cunningham	1st Apollo test—Earth orbit
Apollo	8	December 21, 1968	**Borman, Lovell, Anders**	1st Moon journey—lunar orbits
Apollo	9	March 3, 1969	McDivitt, **Scott**, Schweickart	1st LEM—Earth orbits
Apollo	10	May 18, 1969	**Stafford, Young, Cernan**	1st LEM—lunar orbits
Apollo	11	July 16, 1969	**Armstrong, Collins, Aldrin**	1st Moon landing
Apollo	12	November 14, 1969	**Conrad, Gordon, Bean**	Precision landing
Apollo	13	April 11, 1970	**Lovell, Swigert, Haise**	No landing
Apollo	14	January 31, 1971	**Shepard, Roosa, Mitchell**	Use of MET "rickshaw"
Apollo	15	July 26, 1971	**Scott, Worden, Irwin**	1st use of Rover
Apollo	16	April 16, 1972	**Young, Mattingly, Duke**	22 miles covered
Apollo	17	December 7, 1972	**Cernan, Evans, Schmitt**	Last landing (geologist)
Skylab	1	May 25, 1973	**Conrad**, Kerwin, Weitz	28 days duration
Skylab	2	July 28, 1973	**Bean**, Garriott, Lousma	59 days duration
Skylab	3	November 16, 1973	Carr, Gibson, Pogue	84 days uration
ASTP	—	July 15, 1975	**Stafford**, Brand, Slayton	1st international docking
Shuttle	STS-1	April 12, 1981	**Young**, Crippen	1st Shuttle flight
Shuttle	STS-4	June 27, 1982	**Mattingly**, Hartsfield	Last Shuttle test flight
Shuttle	STS-9	November 28, 1983	**Young**, plus 5 others	1st Spacelab mission
Shuttle	STS-51-C	January 24, 1985	**Mattingly**, plus 4 others	Military satellite deployment

APPENDIX D

Back to the Moon

As you have read, it was a very dangerous and expensive process to achieve the Apollo Moon landings. As a consequence, there has been very little activity related to *going back* to land on the Moon since the end of the Apollo era in 1972, and in particular there have been no further attempts at human landings. When will man be going back to the Moon? Not very soon. The former Soviet Union abandoned its own attempts at landing a human on the Moon after the successes of Apollo, and we know that story via the accounts of Alexei Leonov, the cosmonaut who had been training for that mission.[20.3] Proposals have been published for possible human lunar orbit missions, something akin to Apollo 8, with Elon Musk's company SpaceX offering a circum-lunar space tourism flight using their Dragon capsule, possibly as soon as 2019, and NASA is beginning to consider a Deep Space Gateway concept that would involve humans in lunar orbit around 2022. And presumably, manned landings might eventually follow suit—China is reported to be aiming at a manned lander, but not until the 2030s. It is, therefore, quite probable that by the time that the next human steps on the Moon, there will be no-one left from the original Moon travelers to bear witness.

Starting in about 1992, however, there have been a succession of the less-challenging *robotic* spacecraft placed in lunar orbit by a series of governments: by Japan (Hiten, Selene), by the US (Prospector, LRO, Artemis, Grail, LADEE), by Europe (SMART-1), by China (Chang'e 1, 2, and 3), and by India (Chandrayaan-1). In the Apollo era, the Russians achieved robotic soft landings, and even sample-return with Luna 16, 20, and 24, but there has been only one robotic soft landing since those days, conducted by China with its Yutu rover in December, 2013. It is known that both Japan and India are planning to join the exclusive club (of the US, Russia, and China) who have managed to achieve a lunar soft landing, and China is currently planning further lunar activities with both a lunar far-side rover, and a sample-return mission, and maybe also Russia will be renewing its own sample-return activities. But the human missions are far off into the future. No-one can afford 5% of GDP to repeat the Apollo achievements.

Maybe a totally new approach is needed to get us back to the Moon on a sustainable basis. It needs to have commercial incentives, rather than governmental budgets. Maybe today's college students can find a much cheaper way of doing things, 50 years on from the Apollo era. They have after all given us cubesats and 3-D manufacturing techniques, which offer orders of magnitude benefits in cost savings over more traditional (i.e., governmental) approaches in the satellite business. The author is one of nine international judges for the Google Lunar XPRIZE, which aims to encourage *non-governmental*, commercial, robotic lunar missions via prize incentives. As this book is being written, there are small teams, with many of their members working part-time, from as far apart as Japan, Israel, India, and the US, attempting to soft-land a payload on the Moon, have it travel 500 meters, and then transmit high-definition images back to Earth. This will not be easy to achieve, and so far, the groups have been trying for a decade to get ready to launch. We may recall that, at the start of the space program, it took five failed attempts before the best efforts of the Soviet Union could get a spacecraft to even hit the Moon. And it was equally difficult for the US, which finally made it with Ranger 7, five years after the Soviets had done it with their Luna 2. And it was not until 1966, after many more failures, that the USSR soft-landed Luna 9, and the US landed Surveyor 1. The task was so difficult, even for governments, that China's 2013 rover was the first since the Russian Lunokhod 2 in January, 1973, 40 years earlier.

Maybe, therefore, this new competition amongst non-governmental teams, many of them consisting largely of students, will be the next stage of focus for getting mankind back to the Moon, if they can pull it off. The first XPRIZE was, after all, successful in getting sub-orbital space tourism started, when SpaceShipOne was built and sent into space twice within a two-week period, back in 2004. At the time it was considered to be something only governments could do. These new youthful teams are attempting to win various prizes of the Google Lunar XPRIZE (GLXP) competition. There is a Grand Prize of $20M for the first team to fully complete all the mission requirements. And there are a series of bonus prizes related to achievements which exceed the basic mission, e.g., a range prize for journeys of 5km from the start point. Teams may also compete for lunar night survival and water detection prizes. And finally there are heritage prizes for imagery which shows evidence of any of the prior Moon landings, whether robotic or human. You will be pleased to learn that teams receive guidance about protection and conservation of these artifacts from the earlier historic missions. But won't it be neat to see some of these sites again? Maybe we can find **Al Shepard**'s golf ball! More information about the teams competing for the GLXP—*MoonExpress* and *Synergy Moon* from the US, *Indus* from India, *SpaceIL* from Israel, and *Hakuto* from Japan, can be found at: www.lunar.xprize.org

APPENDIX E

Going Farther

The Apollo approach and architecture took us to the Moon half a century ago. It served the purpose at the time, but was not financially sustainable for ongoing developments. We have not been able to find a financially acceptable way to go back since. How could this be done? It is important to establish the "mode" or main architecture—just as it had been back in the 1960s, when John Houbolt's Lunar Orbit Rendezvous technique was eventually adopted, after facing much criticism, and led to the successful achievement of the Kennedy challenge. What should be the appropriate architecture for 21st-century space

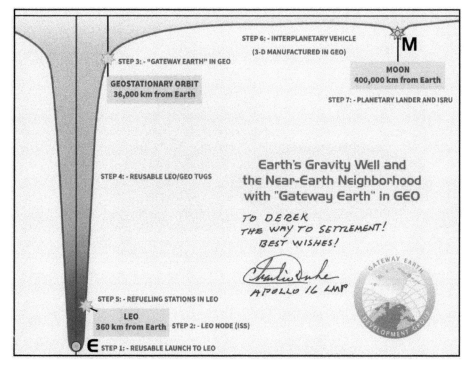

Figure E.1. **Charlie Duke** adds his supportive comment to the Gateway Earth gravity well cross-section chart.

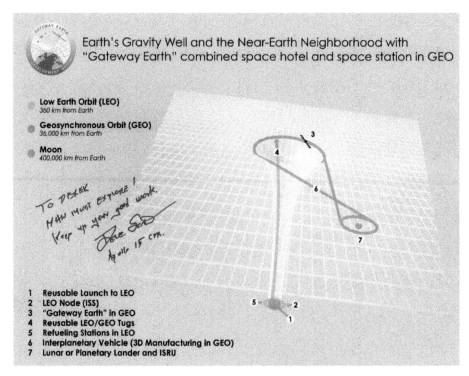

Figure E.2. Dave Scott adds his contribution to this chart of the Gateway Earth concept. "Man must explore," were the words **Scott** used on stepping off the landing pad of Apollo 15's LEM, *Falcon*, in July, 1971.

exploration and development? What about going further across the Solar System? Several Apollo astronauts have suggested that there may have to be a commercial motivation for doing so. The author has proposed one such idea, which uses revenues from space tourism to make it possible. The method goes back to the fundamentals of physics, Newton, and gravity wells to establish a new beginning. For more detail check out *www.GatewayEarth.space*. Meanwhile, here are two summary charts explaining the concept, suitably inscribed by two of the Moon travelers: **Duke** and **Scott** (Figures E.1 and E.2).

The idea is to have a hotel for space tourists placed in geostationary orbit (GEO), which is 100 times farther into space than the location of the ISS which sits in low Earth orbit (LEO). Tourists would visit and stay for about 2 weeks, and they would be able to see an entire hemisphere of the Earth beneath them. The tourists, and their supplies, would go there and back in re-usable space tugs operated by the space tourism operator. Their hotel, and the re-usable tugs, would be paid for from the space tourism revenues. The tugs would be refueled in LEO between excursions up and down to the GEO hotel.

It happens that the geostationary orbit is near the edge of Earth's gravity well. So, any vehicle starting from there would find it a simple, low-energy, and low-cost, matter to proceed to the Moon, Mars, or anywhere else across the Solar System from that starting point.

The proposal, therefore, is to attach a governmental station to the GEO hotel (the combined complex is being called Gateway Earth), and this station would be used to assemble the interplanetary craft intended for going farther— built using 3-D manufacturing techniques. The interplanetary craft can be light-weight, require only small engines, and need no heatshields since they will never have to survive passage through Earth's atmosphere, or launch from Earth's surface. The governmental astronauts would get up and down to Gateway Earth by sharing a taxi ride with the space tourists on their regular journeys up and down to the Gateway Earth complex. This approach is much cheaper than the Apollo Saturn 5 method. Check out the website: *www.GatewayEarth.space* to review progress on the Gateway Earth architecture.

Just as during the early era of space exploration, there needed to be debate, fueled wherever possible with objective data, to decide on the main architecture for moving forward—and in that case, such luminaries as von Braun, Max Faget, and John Houbolt were very vocal in expressing their views and con-cerns—it is now time for a new debate. This time it is about sustainable, long-term space developments. There needs to be a comparison of alternatives, which take into account all of the factors: safety, technological capability, engine design, financial, political, commercial, governmental, and international considerations. It is the hope of the author that the Gateway Earth architecture will be one of the alternatives being considered in such a wide-ranging re-assess-ment of options. Meanwhile, the Gateway Earth Development Group continues to develop, establish the case, and collect the data.

Sources

1 AGGREGATED ASTRONAUT ACCOUNTS

The following 20 sources (1.1 thru 1.20) are excellent accounts, recommended for background research, each covering many, if not all, of the 24 Moon travelers to some degree, and form an aggregate resource for this work. In the 24 subsequent sets of sources (2.1 thru 25.6) I provide the specific main references for each individual Moon traveler.

1.1   Cassutt, M., "Who's Who in Space—the First 25 years," G.K.Hall and Co, 1987.
(A fine data resource for the biographies of the astronauts and cosmonauts during the first 25 years of crewed spaceflight.)

1.2   Chaikin, A., "A Man on the Moon—The Voyages of the Apollo Astronauts," Viking, 1994.
(The best book on the Apollo missions and their astronauts, subsequently used as the basis for a TV mini-series.)

1.3   Chaikin, A. and Kohl, V., "Voices from the Moon—Apollo Astronauts Describe their Lunar Experiences," Viking Penguin, 2009.
(A magnificent book, with superb photographs and a vast store of original quotations from the astronauts. Probably the author's favorite amongst the many books about Apollo.)

1.4   Constantine, M., "Apollo the Panoramas," Moonpans, 2015.
(A superb collection of panoramic photographs showing all the Apollo landing sites, accompanied by comments from some of the Moonwalkers who took the photos.)

1.5   Gibson, E., "The Greatest Adventure," C. Pierson Publishers, 1994.
(Produced by the Association of Space Explorers, this is a handsome work containing excellent photographs illustrating a series of essays by astronauts from all over the world.)

1.6   Harland, D. M., "Exploring the Moon—The Apollo Expeditions," Springer–Praxis, 2008 (second edition).
(An encyclopedic collection of great imagery, both black and white and color, supporting a very detailed account of each Moonwalk.)

1.7   Heiken, G. and Jones, E., "On the Moon—The Apollo Journals", Springer–Praxis, 2007.
(A great source of transcripts of the conversations of the Moonwalkers during their missions, with considerable detail about the activities that took place during each moonwalk.)

1.8   Israelian, G. and May, B., "Starmus," Canopus Publishing, 2014.
(An unusual forum of astronomers, musicians, and astronauts, meeting on the Canary Islands, led to this book, which contains some thoughtful commentary by those astronauts who attended.)

1.9   Kelley, K.W. (Ed), "The Home Planet," Addison Wesley, 1988.
       (When this book came out, in 1988, it was the first of the "coffee table books" published by the Association of Space Explorers, and was stunning in its imagery and the quotations from a truly international mix of astronauts, provided in their original language as well as English translation.)

1.10   Latimer, D., "All We Did was Fly to the Moon," Whispering Eagle, 1975.
       (A small, but essential, basic reference source for all the US spaceflights up to ASTP. The book also includes, for each mission, an account of how the mission patch was designed.)

1.11   Pyle, R., "Destination Moon—The Apollo Missions in the Astronauts' Own Words," Harper Collins, 2005.
       (Another fine illustrated account of the Apollo missions containing quotations by the Moon travelers.)

1.12   Roth, J. (Producer), "The Wonder of it All," DVD/Indian, 2009.
       *(A fine DVD containing face-to-face interviews providing testimony from 7 of the 24 Moon travelers, all of them Moonwalkers—**Aldrin**, **Bean**, **Cernan**, **Duke**, **Mitchell**, **Schmitt**, and **Young**.)*

1.13   Sacknoff, S. (Ed.), "In Their Own Words: Conversations with the Astronauts and Men Who Led America's Journey into Space and to the Moon," Space Publications, 2003.
       (A book of oral histories containing the records of five of the Moon travelers. These interviews were originally published in the magazine "Quest.")

1.14   Sington, D. (Producer), "In the Shadow of the Moon", DVD/DOX, 2007.
       *(Similar to the 1.12 DVD, but this time providing interviews with 10 of the 24 Moon travelers—**Collins**, **Cernan**, **Bean**, **Duke**, **Lovell**, **Mitchell**, **Schmitt**, **Aldrin**, **Scott**, and **Young**.)*

1.15   Slayton, D.K. and Cassutt, M., "Deke!", Forge, 1994.
       (Deke's autobiography, published posthumously, but containing an unrivaled account of how the decisions were made on which astronauts would fly which missions, and how they would be trained.)

1.16   Smith, A., "Moondust," Fourth Estate, 2005.
       (This was the account of a journalist who had first-person meetings with eight of the Moonwalkers.)

1.17   Stewart, M. (Producer), "The Last Man on the Moon," DVD/Stopwatch, 2016.
       *(**Gene Cernan**'s own testamentary DVD, which also contains interview material with **Bean**, **Gordon**, and **Lovell**.)*

1.18   Swanson, G.E. (Ed.), "Before This Decade Is Out …: Personal Reflections on the Apollo Program," University Press of Florida, 2002.
       (A book of oral histories including luminaries such as von Braun, Faget, Kranz, and also two of the Moonwalkers.)

1.19   Wofinger, K. (Producer), "To the Moon," DVD/NOVA, 1999.
       *(A wonderful two-hour documentary covering the whole story from Von Braun through Mercury, Gemini, and Apollo. Contains great interview material with senior NASA management (Seamans, Mueller, etc.) explaining the background to the Kennedy decision process, with ground controllers (Kraft, Kranz, Lunney, Bales, Leibergot, etc.), scientists (Al-Baz, Hartman, etc.), fascinating coverage of the Apollo architecture debate (Faget, Houbolt, etc.), and first-person interviews with 10 of the 24 Moon travelers (**Haise**, **Schmitt**, **Conrad**, **Aldrin**, **Cernan**, **Borman**, **Lovell**, **Anders**, **Stafford**, and **Scott**.)*

1.20   Wolfe, T., "The Right Stuff," 1979.
        *(This wonderful account of the fraternity of space, and his invention of the term "the
        Right Stuff," told about the Mercury astronauts, took Wolfe 6 years to write in book
        form—then it became a great movie, and established permanently the folk lore about
        Al Shepard and Deke Slayton in particular.)*

**Note: for each of the 24 Moon travelers, in alphabetical order, the following
sources provide the specific background material—sometimes re-stating a reference source from the "Aggregated Astronaut Account" list, where some useful
astronaut-specific material was referenced.**

2   ALDRIN

Official NASA/JSC bio, plus:

2.1   Aldrin, B.,"Apollo 11 plus 40 years" Talk at Smithsonian National Air and Space
        Museum, July 19, 2009 (author's recorded cassette tape).
2.2   Aldrin, B., Comments at University Club, Washington DC, meeting of Washington
        Space Business Round-Table, January 16, 2003 (author's recorded cassette tape).
2.3   Aldrin, B., NASA/JSC Oral Histories, July, 1970.
2.4   Aldrin, B., Remarks at "Space Imperatives" Conference to Celebrate Wright
        Brothers Centenary, Reagan Building, Washington DC, December 18, 2003
        (author's recorded cassette tape).
2.5   Aldrin, B., Remarks on Aldrin/Purdue Mars Cycler at International Space
        Development Conference in Toronto, Canada, May 22, 2015 (author's recorded
        cassette tape).
2.6   Aldrin, B., Telephone discussion with author, July 26, 2016.
2.7   Aldrin, B. and Abraham, K., "Magnificent Desolation—The Long Journey Home
        from the Moon", Harmony Books, 2009.
2.8   Aldrin, B. and Abraham, K., "No Dream is too High—Life Lessons from a Man
        Who Walked on the Moon", National Geographic, 2016.
2.9   Aldrin, B. and David, L., "Mission to Mars—My Vision for Space Exploration",
        National Geographic, 2013.
2.10   Aldrin, B. and Warga, W., "Return to Earth," Random House, 1973.
2.11   Armstrong, Aldrin, and Collins, with Gene Farmer and Dora Jane Hamblin, "First
        on the Moon—A Voyage with Neil Armstrong, Michael Collins and Edwin E
        Aldrin Jr", Konecky and Konecky, 1969.
2.12   Chaikin, A. and Kohl, V., "Voices from the Moon: Apollo Astronauts Describe
        Their Lunar Experiences," Viking Penguin, 2009.
2.13   Gibson, E., "The Greatest Adventure," C Pierson Publishers, 1994.
2.14   Israelian, G. and May, B., "Starmus," Canopus Publishing, 2014.
2.15   Kelley, K.W. (Ed), "The Home Planet," Addison Wesley, 1988.
2.16   Roth, J. (Producer), "The Wonder of It All," DVD/Indian, 2009.
2.17   Schorer, L.J., "Kids to Space—A Space Traveler's Guide", Apogee Books, 2006.
2.18   Sington, D. (Producer), "In the Shadow of the Moon", DVD/DOX, 2007.
2.19   Webber, D., "The Wright Stuff—The Century of Effort Behind Your Ticket to
        Space", Apogee Books, 2010.

3 ANDERS

Official NASA/JSC bio, plus:

3.1 Anders, W., Lindbergh Lecture at Smithsonian National Air and Space Museum, May 16, 2007 (author's recorded cassette tape).
3.2 Anders, W., NASA/JSC Oral Histories, October, 1997.
3.3 Anders, W., "40 years after Apollo 8" Talk at Smithsonian National Air and Space Museum, November 13, 2008 (author's recorded cassette tape).
3.4 Chaikin, A. and Kohl, V., "Voices from the Moon: Apollo Astronauts Describe Their Lunar Experiences," Viking Penguin, 2009.
3.5 Israelian, G. and May, B., "Starmus," Canopus Publishing, 2014.

4 ARMSTRONG

Official NASA/JSC bio, plus:

4.1 Armstrong, N.A., "Apollo 11 plus 40 years" Talk at Smithsonian National Air and Space Museum, July 19, 2009 (author's recorded cassette tape).
4.2 Armstrong, N.A., NASA/JSC Oral Histories, February, 1969.
4.3 Armstrong, N.A., NASA/JSC Oral Histories, September, 2001.
4.4 Armstrong, N.A., Presentation at NASA Advisory Council Meeting, Holiday Inn Capitol, Washington, DC, February 8, 2007 (author's recorded cassette tape).
4.5 Armstrong, Aldrin, and Collins, with Gene Farmer and Dora Jane Hamblin, "First on the Moon—A Voyage with Neil Armstrong, Michael Collins and Edwin E. Aldrin Jr.", Konecky and Konecky, 1969.
4.6 Chaikin, A. and Kohl, V., "Voices from the Moon—Apollo Astronauts Describe their Lunar Experiences", Viking Penguin, 2009.
4.7 Hansen, J.R., "First Man—The Life of Neil A. Armstrong", Simon and Schuster, 2005.
4.8 Israelian, G. and May, B., "Starmus," Canopus Publishing, 2014.

5 BEAN

Official NASA/JSC bio, plus:

5.1 Bean, A., "Alan Bean—Artist Astronaut," DVD/JR Productions, 2007.
5.2 Bean, A., Discussion with author (from author's notes taken at the time) at "Space Imperatives" Conference, Ronald Reagan Building, Washington, DC, December 18, 2003.
5.3 Bean, A., Discussion with author (from author's notes taken at the time) at Novaspace book-signing event, November 7, 1998, in Tucson, Arizona.
5.4 Bean, A., NASA/JSC Oral Histories, June, 1998.
5.5 Bean, A., "Painting Apollo—First Artist on Another World", Smithsonian Books, 2009.

5.6   Bean, A., Presentation on 40th Anniversary of Apollo Moon Landings at Smithsonian National Air and Space Museum, November 11, 2009 (author's recorded cassette tape).

5.7   Bean, A. and Chaikin, A., "Apollo: An Eyewitness Account by Astronaut/Explorer Artist/Moonwalker Alan Bean," Greenwich Workshop Press, 1998.

5.8   Chaikin, A. and Kohl, V., "Voices from the Moon—Apollo Astronauts Describe their Lunar Experiences", Viking Penguin, 2009.

5.9   Constantine, M., "Apollo the Panoramas," Moonpans, 2015.

5.10   Gibson, E., "The Greatest Adventure," C Pierson Publishers, 1994.

5.11   Kelley, K.W. (Ed.), "The Home Planet," Addison Wesley, 1988.

5.12   Roth, J. (Producer), "The Wonder of it All," DVD/Indian, 2009.

5.13   Sington, D. (Producer), "In the Shadow of the Moon," DVD/DOX, 2007.

5.14   Stewart, M. (Producer), "The Last Man on the Moon," DVD/Stopwatch, 2016.

6   BORMAN

Official NASA/JSC bio, plus:

6.1   Borman, F., NASA/JSC Oral Histories, April, 1999.

6.2   Borman, F., "40 years after Apollo 8" Talk at Smithsonian National Air and Space Museum, November 13, 2008 (author's recorded cassette tape).

6.3   Borman, F. and Serling, R.J., "Countdown—An Autobiography", Silver Arrow, 1988.

6.4   Chaikin, A. and Kohl, V., "Voices from the Moon—Apollo Astronauts Describe their Lunar Experiences", Viking Penguin, 2009.

6.5   Wofinger, K. (Producer), "To the Moon," DVD/NOVA, 1999.

7   CERNAN

Official NASA/JSC bio, plus:

7.1   Cernan, E., "Apollo 17 plus 30 years"—Wernher von Braun Talk at Smithsonian National Air and Space Museum, March 18, 2003 (author's recorded cassette tape).

7.2   Cernan, E., Discussion with author (from author's notes taken at the time) at "Space Imperatives" Conference, Ronald Reagan Building, Washington, DC, December 18, 2003.

7.3   Cernan, E., NASA/JSC Oral Histories, December, 2007.

7.4   Cernan, E. and Davis, D., "The Last Man on the Moon," St Martin's Press, 1999.

7.5   Chaikin, A. and Kohl, V., "Voices from the Moon—Apollo Astronauts Describe their Lunar Experiences", Viking Penguin, 2009.

7.6   Kelley, K.W. (Ed), "The Home Planet," Addison Wesley, 1988.

7.7   Roth, J. (Producer), "The Wonder of It All," DVD/Indian, 2009.

7.8   Sington, D. (Producer), "In the Shadow of the Moon," DVD/DOX, 2007.

7.9   Stewart, M. (Producer), "The Last Man on the Moon," DVD/Stopwatch, 2016.

7.10   Wofinger, K. (Producer), "To the Moon," DVD/NOVA, 1999.

8 COLLINS

Official NASA/JSC bio, plus:

8.1   Armstrong, Aldrin, and Collins, with Gene Farmer and Dora Jane Hamblin, "First on the Moon—A Voyage with Neil Armstrong, Michael Collins and Edwin E. Aldrin, Jr.", Konecky and Konecky, 1969.
8.2   Chaikin, A. and Kohl, V., "Voices from the Moon—Apollo Astronauts Describe Their Lunar Experiences", Viking Penguin, 2009.
8.3   Collins, M., "Apollo 11 plus 40 years" Talk at Smithsonian National Air and Space Museum, July 19, 2009 (author's recorded cassette tape).
8.4   Collins, M., "Carrying the Fire—An Astronaut's Journeys", Farrar, Straus and Giroux, 1974.
8.5   Collins, M., "Liftoff—The Story of America's Adventure in Space", Grove Press, 1988.
8.6   Collins, M., NASA/JSC Oral Histories, October, 1997.
8.7   Gibson, E., "The Greatest Adventure," C Pierson Publishers, 1994.
8.8   Kelley, K.W. (Ed), "The Home Planet", Addison Wesley, 1988.
8.9   Sington, D. (Producer), "In the Shadow of the Moon," DVD/DOX, 2007.

9 CONRAD

Official NASA/JSC bio, plus:

9.1   Chaikin, A. and Kohl, V., "Voices from the Moon—Apollo Astronauts Describe Their Lunar Experiences", Viking Penguin, 2009.
9.2   Conrad, C., Remarks at Conference of Space Frontier Foundation, Sheraton Hotel, Los Angeles, California, October 1997 (author's personal notes recorded at the event).
9.3   Conrad, N. and Klausner, H.A., "Rocketman—Astronaut Pete Conrad's Incredible Ride to the Moon and Beyond", New American Library, 2006.
9.4   Gibson, E., "The Greatest Adventure," C Pierson Publishers, 1994.
9.5   Sacknoff, S. (Ed.), "In Their Own Words—Conversations with the Astronauts and Men Who Led America's Journey into Space and to the Moon", Space Publications, 2003.

10 DUKE

Official NASA/JSC bio, plus:

10.1   Chaikin, A. and Kohl, V., "Voices from the Moon—Apollo Astronauts Describe their Lunar Experiences", Viking Penguin, 2009.
10.2   Constantine, M., "Apollo the Panoramas," Moonpans, 2015.
10.3   Duke, C., Discussion with author (from author's notes taken at the time) at "Space Imperatives" Conference, Ronald Reagan Building, Washington, DC, December 18, 2003.
10.4   Duke, C., NASA/JSC Oral Histories, March, 1999.

10.5   Duke, C., Remarks on the occasion of the 40th Anniversary of Apollo 11, at the Newseum, Washington DC, July 20, 2009 (author's recorded cassette tape).

10.6   Duke, C. and Duke, D., "Moonwalker—The True Story of an Astronaut Who Found that the Moon wasn't High Enough to Satisfy his Desire for Success", Oliver Nelson, 1990.

10.7   Israelian, G. and May, B., "Starmus," Canopus Publishing, 2014.

10.8   Kelley, K.W. (Ed.), "The Home Planet," Addison Wesley, 1988.

10.9   Roth, J. (Producer), "The Wonder of It All," DVD/Indian, 2009.

10.10   Sacknoff, S. (Ed.), "In their own Words—Conversations with the Astronauts and Men Who Led America's Journey into Space and to the Moon", Space Publications, 2003.

10.11   Sington, D. (Producer), "In the Shadow of the Moon," DVD/DOX, 2007.

10.12   Swanson, G.E. (Ed.), "Before This Decade Is Out ...: Personal Reflections on the Apollo Program," University Press of Florida, 2002.

11 EVANS

Official NASA/JSC bio, plus:

11.1   Chaikin, A. and Kohl, V., "Voices from the Moon—Apollo Astronauts Describe their Lunar Experiences", Viking Penguin, 2009.

11.2   Kelley, K.W. (Ed)., "The Home Planet," Addison Wesley, 1988.

12 GORDON

Official NASA/JSC bio, plus:

12.1   Chaikin, A. and Kohl, V., "Voices from the Moon—Apollo Astronauts Describe their Lunar Experiences", Viking Penguin, 2009.

12.2   Gordon, R., Discussion with author (from author's notes taken at the time) at "Space Imperatives" Conference, Ronald Reagan Building, Washington, DC, December 18, 2003.

12.3   Gordon, R., NASA/JSC Oral Histories, October, 1997.

12.4   Gordon, R.,NASA/JSC Oral Histories, June, 1999.

12.5   Stewart, M. (Producer), "The Last Man on the Moon," DVD/Stopwatch, 2016.

13 HAISE

Official NASA/JSC bio, plus:

13.1   Chaikin, A. and Kohl, V., "Voices from the Moon—Apollo Astronauts Describe their Lunar Experiences", Viking Penguin, 2009.

13.2   Haise, F., NASA/JSC Oral Histories, March, 1999.

13.3   Haise, F., Remarks at Annual John H Glenn Lecture at Smithsonian National Air and Space Museum celebrating 40 years after Apollo 13, April 15, 2010 (from author's notes taken at time).

13.4   Lovell, J. and Kluger, J., "Lost Moon—The Perilous Voyage of Apollo 13", Houghton Mifflin, 1994.

13.5   Sacknoff, S. (Ed.), "In their own Words—Conversations with the Astronauts and Men Who Led America's Journey into Space and to the Moon", Space Publications, 2003.

14 IRWIN

Official NASA/JSC bio, plus:

14.1   Chaikin, A. and Kohl, V., "Voices from the Moon—Apollo Astronauts Describe their Lunar Experiences", Viking Penguin, 2009.

14.2   Gibson, E., "The Greatest Adventure," C Pierson Publishers, 1994.

14.3   Irwin, J.B. and Emerson, W.A. "To Rule the Night—The Discovery Voyage of Astronaut Jim Irwin", A J Holman Company, 1973.

14.4   Kelley, K.W. (Ed.), "The Home Planet," Addison Wesley, 1988.

15 LOVELL

Official NASA/JSC bio, plus:

15.1   Chaikin, A. and Kohl, V., "Voices from the Moon—Apollo Astronauts Describe their Lunar Experiences", Viking Penguin, 2009.

15.2   Dick, S.J. and Cowing, K.L. (Eds.), "Risk and Exploration—Earth, Sea and the Stars", NASA, 2004.

15.3   Gibson, E., "The Greatest Adventure," C Pierson Publishers, 1994.

15.4   Israelian, G. and May, B., "Starmus," Canopus Publishing, 2014.

15.5   Kelley, K.W. (Ed.), "The Home Planet," Addison Wesley, 1988.

15.6   Lovell, J., Remarks at Annual John H Glenn Lecture at Smithsonian National Air and Space Museum celebrating 40 years after Apollo 13, April 15, 2010 (from author's notes taken at time).

15.7   Lovell, J., NASA/JSC Oral Histories, May, 1999.

15.8   Lovell, J., "40 yrs after Apollo 8" Talk at Smithsonian National Air and Space Museum, November 13, 2008 (author's recorded cassette tape).

15.9   Lovell, J. and Kluger, J., "Lost Moon—The Perilous Voyage of Apollo 13", Houghton Mifflin, 1994.

15.10   Sacknoff, S. (Ed.), "In their own Words—Conversations with the Astronauts and Men Who Led America's Journey into Space and to the Moon", Space Publications, 2003.

15.11   Sington, D. (Producer), "In the Shadow of the Moon," DVD/DOX, 2007.

15.12   Stewart, M. (Producer), "The Last Man on the Moon," DVD/Stopwatch, 2016.

16　MATTINGLY

Official NASA/JSC bio, plus:

16.1　Chaikin, A. and Kohl, V., "Voices from the Moon—Apollo Astronauts Describe their Lunar Experiences", Viking Penguin, 2009.
16.2　Dick, S.J. and Cowing, K.L. (Eds.), "Risk and Exploration—Earth, Sea and the Stars", NASA, 2004.
16.3　Kelley, K.W. (Ed), "The Home Planet," Addison Wesley, 1988.
16.4　Mattingly, T.K., NASA/JSC Oral Histories, November, 2001.
16.5　Mattingly, T.K., NASA/JSC Oral Histories, April, 2002.
16.6　Mattingly, T.K., Remarks at Annual John H Glenn Lecture at Smithsonian National Air and Space Museum celebrating 40 years after Apollo 13, April 15, 2010 (from author's notes taken at the time).
16.7　Mattingly, T.K., Remarks at the Space Frontier Conference, Hollywood, California, October 19, 1996 (from author's notes taken at the time).

17　MITCHELL

Official NASA/JSC bio, plus:

17.1　Chaikin, A. and Kohl, V., "Voices from the Moon—Apollo Astronauts Describe their Lunar Experiences", Viking Penguin, 2009.
17.2　Constantine, M., "Apollo the Panoramas," Moonpans, 2015.
17.3　Kelley, K.W. (Ed.), "The Home Planet," Addison Wesley, 1988.
17.4　Krone, B. (Ed.), "Beyond Earth—The Future of Humans in Space", Apogee Books, 2006.
17.5　Mitchell, E., Discussion at "Overview Effect" Conference, Doubletree Hotel, Arlington, Virginia, July 18, 2007 (from author's notes taken at the time).
17.6　Mitchell, E., Discussion with author (from author's notes taken at the time) at the United Societies in Space Convention, at the Warwick Hotel, Denver, Colorado, August 3, 1997.
17.7　Mitchell, E., "Earthrise: My Adventures as an Apollo 14 Astronaut," Chicago Review Press, 2014.
17.8　Mitchell, E., Email exchange with author, August 27, 2006.
17.9　Mitchell, E., Email exchange with author, October 18, 2012.
17.10　Mitchell, E., NASA/JSC Oral Histories, September, 1997.
17.11　Mitchell, E., "The View from Space—A Message of Peace," DVD/SMPI, 2006.
17.12　Mitchell, E. and Williams, D., "The Way of the Explorer—An Apollo Astronaut's Journey Through the Material and Mystical Worlds", Putnam, 1996.
17.13　Roth, J. (Producer), "The Wonder of It All," DVD/Indian, 2009.
17.14　Sington, D. (Producer), "In the Shadow of the Moon," DVD/DOX, 2007.

18 ROOSA

Official NASA/JSC bio, plus:

18.1   Chaikin, A. and Kohl, V., "Voices from the Moon—Apollo Astronauts Describe their Lunar Experiences", Viking Penguin, 2009.

18.2   Moseley, W.G., "Smoke Jumper, Moon Pilot—The Remarkable Life of Apollo 14 Astronaut Stuart Roosa", Acclaim Press, 2011.

19 SCHMITT

Official NASA/JSC bio, plus:

19.1   Chaikin, A. and Kohl, V., "Voices from the Moon—Apollo Astronauts Describe their Lunar Experiences", Viking Penguin, 2009.

19.2   Constantine, M., "Apollo the Panoramas," Moonpans, 2015.

19.3   Dick, S.J. and Cowing, K.L. (Eds.), "Risk and Exploration—Earth, Sea and the Stars", NASA, 2004.

19.4   Gibson, E., "The Greatest Adventure," C Pierson Publishers, 1994.

19.5   Roth, J. (Producer), "The Wonder of It All," DVD/Indian, 2009.

19.6   Schmitt, H.H., "Apollo 17 plus 30 years" Annual Wernher von Braun talk at Smithsonian National Air and Space Museum, March 18, 2003 (author's recorded cassette tape).

19.7   Schmitt, H.H., Comments at NASA Advisory Council Meeting, Holiday Inn Capitol, Washington, DC, February 8, 2007 (author's recorded cassette tape).

19.8   Schmitt, H.H., Discussion with author (from author's notes taken at the time) at the STAIF Conference, Hilton Hotel, Albuquerque, New Mexico, April 26, 1998.

19.9   Schmitt, H.H., Fellow speaker at "Wright Centenary" Conference in Ronald Reagan Building, Washington, DC, on December 18, 2003 (from author's notes taken at the time).

19.10   Schmitt, H.H., NASA/JSC Oral Histories, July, 1999.

19.11   Schmitt, H.H., NASA/JSC Oral Histories, March, 2000.

19.12   Schmitt, H.H., "Return to the Moon—Exploration, Enterprise and Energy in the Human Settlement of Space", Copernicus Books, 2006.

19.13   Sington, D. (Producer), "In the Shadow of the Moon," DVD/DOX, 2007.

19.14   Swanson, G.E. (Ed.), "Before this Decade is Out …Personal Reflections on the Apollo Program", University Press of Florida, 2002.

20 SCOTT

Official NASA/JSC bio, plus:

20.1   Chaikin, A. and Kohl, V., "Voices from the Moon—Apollo Astronauts Describe their Lunar Experiences", Viking Penguin, 2009.

20.2   Scott, D., Discussion with author (from author's notes taken at the time) at a Convention at Crowne Plaza Hotel, Secaucus, New Jersey, August 13, 2005.

20.3   Scott, D., Leonov, A., and Toomey, C., "Two Sides of the Moon," Thomas Dunne Books, 2004.

20.4   Sington, D. (Producer), "In the Shadow of the Moon," DVD/DOX, 2007.

21   SHEPARD

Official NASA/JSC bio, plus:

21.1   Chaikin, A. and Kohl, V., "Voices from the Moon—Apollo Astronauts Describe their Lunar Experiences", Viking Penguin, 2009.

21.2   Sacknoff, S. (Ed.), "In their own Words—Conversations with the Astronauts and Men Who Led America's Journey into Space and to the Moon", Space Publications, 2003.

21.3   Shepard, A., NASA/JSC Oral Histories, February, 1998.

21.4   Shepard, A., Remarks at "Shoot for the Moon" event at San Diego Aerospace Museum, August 3, 1995 (from author's notes taken at the event).

21.5   Shepard, A., Slayton, D.K., Barbree, J., and Benedict, H., "Moon Shot—The Inside Story of America's Race to the Moon", Turner Publishing, 1994.

21.6   The Mercury Astronauts, "We Seven", Simon and Schuster, 1962.

21.7   Thompson, N., "Light this Candle: The Life & Times of Alan Shepard, America's First Spaceman", Crown Publishing, 2004.

22   STAFFORD

Official NASA/JSC bio, plus:

22.1   Chaikin, A. and Kohl, V., "Voices from the Moon—Apollo Astronauts Describe their Lunar Experiences", Viking Penguin, 2009.

22.2   Gibson, E., "The Greatest Adventure," C Pierson Publishers, 1994.

22.3   Kelley, K.W. (Ed), "The Home Planet," Addison Wesley, 1988.

22.4   Stafford, T.P., "Flight Jacket Night" with Cernan and Schirra, Smithsonian National Air and Space Museum, November 3, 2006 (author's recorded cassette tape).

22.5   Stafford, T.P., NASA/JSC Oral Histories, October, 1997.

22.6   Stafford, T.P., NASA/JSC Oral Histories, April, 2015.

22.7   Stafford, T.P., Talk at book-signing at Barnes and Noble in Washington, DC (author's recorded cassette tape), November 4, 2002.

22.8   Stafford, T.P., Testimony at "Moon, Mars and Beyond" Commission hearing, Department of Commerce, Washington, DC, February 11, 2004 (from author's notes taken at the time).

22.9   Stafford, T.P., 30-year Celebration of ASTP Mission, Smithsonian National Air and Space Museum, July 14, 2005 (from author's notes taken at the time).

22.10   Stafford, T.P. and Cassutt, M., "We Have Capture: Tom Stafford and the Space Race," Smithsonian Books, 2002.

23 SWIGERT

Official NASA/JSC bio, plus:

23.1   Lovell, J. and Kluger, J., "Lost Moon—The Perilous Voyage of Apollo 13", Houghton Mifflin, 1994.

24 WORDEN

Official NASA/JSC bio, plus:

24.1   Chaikin, A. and Kohl, V., "Voices from the Moon—Apollo Astronauts Describe their Lunar Experiences", Viking Penguin, 2009.
24.2   Kelley, K.W. (Ed.), "The Home Planet," Addison Wesley, 1988.
24.3   Worden, A.M., "Hello Earth—Greetings from Endeavour", Nash Publishing, 1974.
24.4   Worden, A.M., NASA/JSC Oral Histories, May, 2000.
24.5   Worden, A.M., Talk given "An Evening with Al Worden" as part of the Re-inventing Space Conference, Bodleian Library Divinity School, Oxford, UK, November 12, 2015 (from author's notes taken at the time).
24.6   Worden, A.M., Talk given "Meet Astronaut Al Worden" at Vandiver Inn, Havre de Grace, Maryland on June 14, 2012 (author's recorded cassette tape).
24.7   Worden, A.M. and French, F., "Falling to Earth—An Apollo 15 Astronaut's Journey to the Moon", Smithsonian Books, 2011.

25 YOUNG

Official NASA/JSC bio, plus:

25.1   Chaikin, A. and Kohl, V., "Voices from the Moon—Apollo Astronauts Describe their Lunar Experiences", Viking Penguin, 2009.
25.2   Grissom, V.I., "Gemini: a Personal Account of Man's Venture into Space," Macmillan, 1968.
25.3   Roth, J. (Producer), "The Wonder of it All," DVD/Indian, 2009.
25.4   Sington, D. (Producer), "In the Shadow of the Moon," DVD/DOX, 2007.
25.5   Young, J.W., Lecture "Spaceflight: Project Gemini to the Space Shuttle" given at Smithsonian National Air and Space Museum, April 11, 2003 (author's recorded cassette tape).
25.6   Young, J.W. and Hansen, J.R., "Forever Young—A Life of Adventure in Air and Space", University Press of Florida, 2012.

Index

Other titles published by Curtis Press

This is a book about the business of space. It is indeed the first such book to explore the creation of the whole new field of commercial space exploration, previously considered to be an oxymoron. The book anchors the story on historical developments which transformed the original governmental and military space program into the successful satellite communications and broadcasting business. It then builds the case for moving forward to the next stages leading to full commercial space exploration, with space tourism and therefore commercial space transportation being key to this paradigm shift. Originally the commercialization of space was limited to various kinds of satellites, but now the march of history is bringing a manned (or rather "crewed") spaceflight regime into its remit. The author has been involved in all these phases of development as a business insider, and was one of the international judges of The Google Lunar XPRIZE which is using prizes to kickstart lunar commercial business.

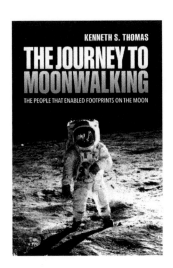

This book represents over two decades of research, interviewing original participants and working with other spacesuit historians to determine Apollo spacesuit contributions and contributors. The author brings a unique expertise to this historic achievement. He was a spacesuit engineer for 22 years and has been a consultant to national museums since 1993. Additionally, performing knowledge capture for NASA gave the author views into the Apollo history at a micro-level, which provided him with further enlightenment. The result is a human chronicle of the challenges, achievements, and experiences related to the most watched historical event of its time.